BIRKHÄUSER

Modeling and Simulation in Science, Engineering and Technology

Nicola Bellomo
Bertrand Lods
Roberto Revelli
Luca Ridolfi

Generalized Collocation Methods

Solutions to Nonlinear Problems

Birkhäuser
Boston • Basel • Berlin

Nicola Bellomo
Dipartimento di Matematica
Politecnico di Torino
Corso Duca degli Abruzzi 24
10129 Torino
Italia

Bertrand Lods
Laboratoire de Mathématiques
Université Blaise Pascal
Campus Universitaire des Cézeaux
63177 Aubière cedex
France

Roberto Revelli
Dipartimento di Idraulica, Trasporti,
ed Infrastrutture Civili
Politecnico di Torino
Corso Duca degli Abruzzi 24
10129 Torino
Italia

Luca Ridolfi
Dipartimento di Idraulica, Trasporti,
ed Infrastrutture Civili
Politecnico di Torino
Corso Duca degli Abruzzi 24
10129 Torino
Italia

Mathematics Subject Classification: 00A72, 74S25, 74S30

Library of Congress Control Number: 2007932360

ISBN-13: 978-0-8176-4525-0 e-ISBN-13: 978-0-8176-4610-3

Printed on acid-free paper.

9 8 7 6 5 4 3 2 1

www.birkhauser.com

Contents

Preface

Nonlinear Problems in Applied Sciences

This book has been proposed to offer engineers and scientists various mathematical tools, based on generalized collocation methods, to solve nonlinear problems related to partial differential and integro-differential equations. This preface first describes the aims of the book, then outlines its contents, and finally develops a critical analysis focused on the merits (and limits) of the method herein described compared to classical applied mathematics approaches.

Aims and Users

Mathematical problems of interest in technology and in applied sciences are often characterized by nonlinear features which may refer either to the model or to the mathematical problems, and, in some cases, to both. Linearity should be regarded as a special case that is often generated by an artificial simplification of a physical reality.

Mathematical models are already an approximation of a physical reality, which is useful in applied sciences although scientists occasionally even refer to falsification of a physical reality. Linearity assumptions generally increase the gap between the description delivered by a model and the effective behavior of the real system. Therefore, dealing with nonlinear problems can contribute to a fruitful interaction between mathematics and applied sciences.

This reasoning has motivated the search for mathematical methods to deal with nonlinear problems generated by the application of models, stated by partial differential and integro-differential equations, to the analysis of real-world problems. This book specifically deals with the development and application of generalized collocation methods—originally called differential quadrature methods.

The application of these methods is based on the approximation and interpolation of the dependent variables by using suitable polynomials or functions according to their values in the collocation points corresponding to a suitable discretization of the space variable. Lagrange polynomials and sinc functions are usually adopted for the interpolation.

Then the space derivatives, or integral terms, are approximated using interpolation. Replacing them in the evolution equation transforms the initial-boundary value problem into an initial value problem for ordinary differential equations that describe the evolution of the values of the dependent variables in the nodes. Boundary conditions are imposed in the collocation points, corresponding to the boundary of the domain of the independent variables. The solution of the initial-boundary value problem is obtained by solving the initial value problem and then interpolating the solution again.

This book has been written both for scientists and engineers who are interested in modelling real systems by using differential or operator equations, and for university students who have a good knowledge of fundamental mathematics and differential calculus, at a masters course level, and are interested in the application of mathematics to technology and applied sciences.

The aim is to offer an easy-to-use handbook for the implementation of the method and for the development and application of scientific programs. This book deals with mathematical applications and is not addressed to mathematicians interested in conceptual topics and proof theory in general. This book is founded on the idea that ***modelling, mathematical methods, and scientific computation should be dealt with together in a unified presentation***. All the above-mentioned features are part of applied mathematics, and a unified presentation can contribute to a deeper understanding of the subjects.

Scientific programs have been proposed in such a way that the reader can find guidelines for the practical application of the method. The programs make use of the software ***Mathematica***®, which offers a friendly approach to programming, as well as a careful optimization of routines related to the application of algorithms. However, the choice of this software is also due to the authors' personal taste; the reader can use other software depending on his/her own experience.

Contents

The contents of the book are organized in two parts. The first part, consisting of the first two chapters, deals with introductory topics concerning a variety of mathematical models, the related statement of mathematical problems, and general aspects of interpolation techniques.

• Chapter 1 concerns modelling and analytical aspects. The first part of the chapter describes a number of mathematical models, generally nonlinear, which will be used to test the application of the mathematical method at a practical level. This chapter also deals with the statement of the problems obtained by implementing the model with suitable initial and boundary conditions.

• Chapter 2 discusses the technical aspects and the scientific programs concerning the interpolation of functions and surfaces by generalized collocation methods. An analysis is made to show how the selection of the interpolating polynomials and their collocation can be technically organized to obtain the best approximation.

The second part, consisting of Chapters 3, 4, 5, and 6, supported by the Appendix, deals with the development of generalized collocation methods to solve a variety of problems related to nonlinear partial differential equations.

• Chapter 3 deals with the solution of initial-boundary value problems for nonlinear partial differential equations in unbounded domains. Mathematical problems are obtained by assigning the initial conditions, and interpolations are developed using sinc functions.

• Chapter 4 concerns the generalization of the method proposed in Chapter 3 to the solution of initial-boundary value problems in one space dimension. The various problems dealt with in Section 1.3 are technically treated referring to the classical Dirichlet, Neumann, and Robin boundary conditions.

• Chapter 5 discusses the application of generalized collocation methods to the solution of initial-boundary value problems in two space dimensions. The contents are a development of the concepts already proposed in the preceding chapters. The application, however, has to deal with an increased computational difficulty.

• Chapter 6 develops various generalizations of the methods to the solution of some ill-posed problems. Specifically, the chapter addresses problems with nonlinear boundary conditions and problems where the boundary conditions, or source terms, are replaced by additional information on the solution of the mathematical problem. These problems are generated by several interesting engineering applications, where at the boundaries measurements can provide only information on nonlinear functions of the dependent variable rather than a direct measurement of Dirichlet or Neumann boundary conditions. This section also discusses the solution of the problems for integro-differential equations and with a critical analysis focused on the validity of the method with respect to other mathematical tools.

• The Appendix reports the various scientific programs used for the applications proposed in this book. This chapter is not simply a collection of Notebooks, but is also a guide to scientific programming. A useful guide for the application of ***Mathematica*** to the solution of problems of interest in applied sciences can be obtained from the books by Bellomo, Preziosi, and Romano (2000), Romano, Lancellotta, and Marasco (2005), and Lynch

(2007).

All the chapters include several examples and problems that have been solved by computational methods. Full details of the solution technique are given, while the last part of each chapter provides problems and exercises so the reader can practice using the mathematical tools of each chapter.

Critical Analysis

Dealing with nonlinear models and problems in applied sciences is a crucial passage of the application of mathematics to real-world analysis. The mathematical methods developed in this book are also proposed considering their immediate ability to deal with nonlinearities.

The analyses of inverse problems and of initial-boundary value problems with nonlinear boundary conditions are particularly important. This specific feature makes the method interesting for engineering applications as also documented in the valuable book by Chang Shu (1992) *Differential Quadrature and Its Application in Engineering*. Our book is directed towards dealing with mathematical foundations of the method with special attention to computational treatments of nonlinearities. Programming with ***Mathematica*** leads to a rapid and efficient production of scientific computation.

However, it would be naive to try to hide some of the well identified limits of the method. While it efficiently deals with nonlinearity, it suffers from computational complexity problems induced by geometrical features, e.g., problems in more than two space variables, especially when they are defined over complex shapes. Additional difficulties are also generated by problems characterized by oscillating solutions, with high frequency, in the space variables.

Applied scientists will be made aware throughout of the the limits (not only of the advantages) of the effective applicability of generalized collocation methods with respect to other methods of applied mathematics.

Finally, let us stress that this book has also been written for engineering schools or applied sciences faculties, as part of courses of applied mathematics where modelling aspects are followed by simulations suitable for providing a deeper understanding of the mathematical model. Because of these reasons, we hope that the book will help in the research activity of those engineers and applied scientists who are involved in mathematical problems where the interplay between mathematics technology, or applied sciences in general, is not trivial.

N. Bellomo, B. Lods, R. Revelli, and L. Ridolfi

1

Mathematical Models and Problems in Applied Sciences

1.1 Introduction

This book deals with computational methods for the solution and simulation of nonlinear evolution problems referred to models of interest in technology and, more generally, in applied sciences.

Differential quadrature and ***collocation methods*** have been developed and applied to a variety of mathematical problems with the aim of providing engineers and applied scientists with all the technical tools (including scientific programs) they need to apply these methods to the solution of various nonlinear mathematical problems.

This first chapter provides a concise introduction to some models and problems in applied sciences as well as to some preliminary guidelines for applying ***generalized collocation methods*** to the solution of mathematical problems for partial differential and integro-differential equations. Problems are generated by the application of mathematical models to the simulation of real-world phenomena.

The contents are not limited to technical applications. The results of computations are often interpreted from a physics and engineering sciences viewpoint, and simulations are used to visualize phenomena which are not always fully observable in experiments.

Scientific programs have been developed using the software ***Mathematica***®, which has the great advantage of using simple programming rules based on the use of optimized subroutines: this feature makes this software suitable for user-friendly scientific programs.

The contents of this first chapter are proposed in four sections.

• Section 1.2 deals with the *description of some models*, of particular interest in applied sciences, described by nonlinear partial differential or integro-differential equations. The selection of these models has been proposed bearing in mind the applications developed in the subsequent chapters.

• Section 1.3 deals with a *dimensional analysis* which is suitable for writing models with dimensionless independent and dependent variables. This technical issue plays a crucial role in the development of computational schemes.

• Section 1.4 provides a brief introduction to the *concept of a mathematical problem* which is obtained by linking the model (or equation) to suitable initial and/or boundary conditions. The statement of the problem is related to a classification of partial differential equations, which in turn is related to qualitative properties of the mathematical models.

• Section 1.5 is a *brief introduction to the guidelines of* ***generalized collocation methods***. It is only a concise outlining of the sequential steps to be followed in the application of the method. Each step will be properly developed in the chapters which follow. The same section offers a *critical analysis* that anticipates some of the advantages and drawbacks of the method; it actually provides an efficient tool for problems in one or two space dimensions, but technical computational difficulties arise in the case of space dimensions larger than three.

Note that although various models are proposed in this chapter, the book does not deal exhaustively with modelling aspects. The reader can refer to the books (among others) by Lin and Segel (1988), Bellomo and Preziosi (1996), and Fowler (1997), to increase his/her knowledge of modelling, the statement of problems, and the mathematical methods for nonlinear evolution problems.

1.2 Some Models in Applied Sciences

This section presents some models of interest in applied sciences which will be used in the subsequent chapters to describe the application of the generalized collocation method. Each model will be used for specific applications.

Referring to the contents of the subsequent chapters, the analysis of problems in unbounded domains is related to linear and nonlinear wave motion models, while initial-boundary value problems are related to hydrodynamic vehicular traffic flow models, to convection-diffusion models, and to reaction-diffusion models. The application of the method to the solution of integro-differential equations is applied to a class of models of population dynamics with an internal structure.

The definitions that follow may be useful not only to help to describe the models, but also for a better understanding of the mathematical problems that can be treated with the methods proposed in this book.

Real-world systems can be observed and phenomenologically studied in order to obtain as detailed a knowledge as possible on the inner structure of the system. This systematic observation can then be developed in the design of a ***mathematical model***, which is an equation, or a set of equations, whose solution (that is, the solution to the related mathematical problems) provides the evolution of the ***variable*** that describes the physical system which is the subject of the modelling process.

Some technical definitions are given according to Bellomo and Preziosi (1996).

• ***Independent variables***: A physical system can be observed in a time interval $[0, T] \subseteq \mathbb{R}_+$ and in a volume $\mathcal{D} \subseteq \mathbb{R}^3$. Therefore, the ***time***

$$t \in [0, T] \subseteq \mathbb{R}_+ \tag{1.2.1}$$

and ***space***

$$\mathbf{x} = \{x, y, z\} \in \mathcal{D} \subseteq \mathbb{R}^3 \tag{1.2.2}$$

are the ***independent variables***.

• The ***state variable***, which acts as a vector-valued ***dependent variable***, is generally a function of the independent variables,

$$\mathbf{u} = \mathbf{u}(t, \mathbf{x}) \, : \, [0, T] \times \mathcal{D} \longrightarrow \mathbb{R}^n \,, \tag{1.2.3}$$

where

$$\mathbf{u} = \{u_1, \ldots, u_i, \ldots, u_n\} \,, \tag{1.2.4}$$

so that $\mathbf{u}$ is the set of the variables that describe, in the mathematical model, the physical state of the real system.

• ***Parameters*** are quantities that characterize the physical system to be modelled. These quantities can be defined either in a dimensional or in a dimensionless form and can be obtained either by direct measurements on the system itself, or by comparisons between the predictions of the model and empirical data referred to the behavior of the real system.

• A mathematical model, defined by a system of equations, is said to be ***consistent*** if the number of linearly independent equations, which describe the evolution of the state variable $\mathbf{u}$, is equal to the dimension of the state-dependent variable.

In order to define the physical volume occupied by the real system, a fixed system of orthogonal axes $O(xyz)$ with unit vectors $\mathbf{i}$, $\mathbf{j}$, and $\mathbf{k}$ can be used. The ***past*** is represented by the negative values of the time, $t \in \mathbb{R}_-$; the ***future*** by positive values $t \in \mathbb{R}_+$. A system can be observed for positive times, however, the mathematical model can also refer to negative times. In this case $t \in \mathbb{R}$.

Both the observation and simulation of systems of the real world need a definition of suitable observation and modelling scales. Different models and descriptions correspond to different representation scales. For instance, if the motion of a fluid in a pipe is observed at a microscopic scale, each particle is observed singularly. Consequently, the motion can be described, in the framework of Newtonian mechanics, by ordinary differential equations which relate the force applied to each particle to its mass times the acceleration. Applied forces are generated by an external field and by interactions among the particles. On the other hand, the same system can be observed and described at a larger scale considering suitable local averages of the mechanical quantities linked to a large number of particles. This means that the model refers to macroscopic quantities, such as mass density and velocity of the fluid. Mass density is the ratio between the mass of the particles contained in a reference volume and the volume itself which is, in principle, as small as possible. A similar definition can be given for the mass velocity, namely the ratio between the momentum of the particles in the reference volume and their mass. Both quantities can be measured by suitable experimental devices operating at a scale of a greater order than that of the single particle.

Bearing all this in mind, the following additional definitions can be given.

• ***Microscopic scale:*** A real system can be observed, measured, and modelled at a ***microscopic scale*** if all the single objects that compose the system are individually considered. The model describes the evolution of the state of each object considered as a whole.

• ***Macroscopic scale:*** A real system can be observed, measured, and modelled at a ***macroscopic scale*** if suitable locally averaged quantities related to the objects that compose the system are considered. The model describes the evolution of the above-mentioned quantities.

• ***Mesoscopic scale:*** A real system can be observed, measured, and modelled at a ***mesoscopic (kinetic) scale*** if it is composed of a large number of interacting objects and the model describes the probability distribution over the state of the objects, and the macroscopic observable quantities related to the system can be recovered from moments weighted by the distribution function of the state of the system.

Microscopic models are generally stated in terms of ordinary differential equations while ***macroscopic models*** are usually stated in terms of partial differential equations, so that the dependent (state) variable corresponds to macroscopic observable quantities.

The contents of the following chapters essentially refer to simulation problems of macroscopic models. The application of the collocation method to the study of integro-differential equations is also developed with reference to the ***mesoscopic*** representation.

Some specific models are described which should be regarded as simple examples, selected for tutorial purposes on the basis of the personal tastes of the authors, to show the practical application of the method and of the computational schemes. Other models can also serve the same purpose, as indicated in the various problems suggested at the end of the chapters devoted to specific applications.

Example 1.2.1
Transport and Diffusion Model

Consider the convection-diffusion of a passive chemical in a stream. The following simple model describes this phenomenon:

$$\frac{\partial u}{\partial t} + c\,\frac{\partial u}{\partial x} = K\,\frac{\partial^2 u}{\partial x^2}\,, \tag{1.2.5}$$

where the state (*dependent*) variable is the concentration u of the chemical. The stream is schematized as a one-dimensional flow. The longitudinal position along the stream is defined by the (*independent*) space variable x. Therefore, the state variable depends on the time t and on the space variable x.

The derivation of this model is obtained by the mass conservation equation for a fluid in a one space dimensional channel:

$$\frac{\partial u}{\partial t} + \frac{\partial q}{\partial x} = 0\,,$$

where u is the mass density and $q = q(t,x)$ is the flow. The above conservation equation is not yet a self-consistent model unless it is linked to a phenomenological model of the material behavior referring q to u.

Model (1.2.5) is obtained using the following expression:

$$q = c\,u - K\,\frac{\partial u}{\partial x}\,,$$

known also as Fick's law, which relates the flow to a transport and to a diffusion term. Therefore, the physical system is obviously influenced by a parameter c, which defines a constant convective velocity, and by a parameter K, which defines the dispersion of u along the space variable.

The above mathematical model can be generalized to the case of a dispersion parameter that depends on the concentration u. In this case, the model writes

$$\frac{\partial u}{\partial t} + c\,\frac{\partial u}{\partial x} = \frac{\partial}{\partial x}\left[K(u)\,\frac{\partial u}{\partial x}\right] \cdot \tag{1.2.6}$$

In some cases, the convective term can also depend on the space variable or even on the variable u. On the other hand, if $c = 0$, the nonlinear diffusion equation:

$$\frac{\partial u}{\partial t} = \frac{\partial}{\partial x}\left[K(u)\frac{\partial u}{\partial x}\right] \cdot \tag{1.2.7}$$

is obtained. This model is linear if $K = K_0$ is a constant.

□

Example 1.2.2.a
Korteweg–deVries Solitary Wave Model

Solitary wave phenomena can be described by the model derived by Korteweg and de Vries in 1895 (KdV for short). In the original formulation it describes the evolution of the long water waves in rectangular cross-sectional channels. The same model is also used in several fluid mechanics fields, such as internal gravity waves in stratified fluids or waves in a rotating atmosphere. The model is characterized by an appropriate balance between nonlinearity and dispersion and can be regarded as one of the relevant paradigms of nonlinear waves and soliton solutions, as documented in, e.g., Whitham (1974) and Johnson (1997).

The KdV model is written as follows:

$$\frac{\partial u}{\partial t} + c_0\left(\frac{\partial u}{\partial x} + \frac{3}{2}\,\frac{u}{d}\,\frac{\partial u}{\partial x} + \frac{1}{6}\,\frac{\partial^3 u}{\partial x^3}\right) = 0, \tag{1.2.8}$$

where x is the longitudinal coordinate, t is the time, $u = u(x,t)$ is the free surface displacement from the equilibrium level of a nonviscous incompressible fluid of quiescent depth d, and $c_0 = \sqrt{gd}$, where g is the gravity.

Let us now define the following dimensionless quantity:

$$\widehat{x} = \frac{x - (1+\alpha)c_0 t}{\ell}\sqrt{U}\,, \qquad \widehat{t} = \frac{d^2 c_0 t \sqrt{U^3}}{6\ell^3}\,,$$

$$\widehat{u} = \frac{3u - 2\alpha d}{a}\,, \qquad U = \frac{3a\ell^2}{d^3}\,,$$

where $a \ll d$ is an amplitude scale, $\ell \gg d$ is a horizontal length scale, and $\alpha = O(a/d)$ is an arbitrary constant.

If the previous dimensionless quantities are introduced into (1.2.8) and the carets dropped, the dimensionless KdV model is obtained:

$$\frac{\partial u}{\partial t} + u\,\frac{\partial u}{\partial x} + \frac{\partial^3 u}{\partial x^3} = 0\,\cdot \tag{1.2.9}$$

The KdV equation can be derived starting from Saint Venant equations, see Henderson (1966),

$$\frac{\partial h}{\partial t} + \frac{\partial (vh)}{\partial x} = 0\,,$$

and

$$\frac{\partial v}{\partial t} + v\frac{\partial v}{\partial x} + g\frac{\partial h}{\partial x} = 0\,,$$

that model one-dimensional nonlinear water waves over a horizontal bottom, where h is the water depth, v is the bulk velocity, and g is gravity acceleration.

Water waves modelled by Saint-Venant equations are hyperbolic and do not model dispersion effects described by the relation: $\omega^2 = gh_0k^2 = c_0^2k^2$, where h_0 is the equilibrium basic state, k is the wave-number, and $c_0 = \sqrt{gh_0}$ is the wave celerity. On the other hand, it is possible to prove, e.g., Whitham (1974), that one-dimensional dispersive water waves are characterized by the dispersion relation $\omega^2 = gk\tanh kh_0$, whose first term of its Taylor expansion corresponds to the dispersion relation of Saint-Venant equations. This observation suggests that a way to describe nonlinear dispersive water waves consists in modifying the Saint-Venant model to obtain a new wave model whose dispersion relation includes the second term of the Taylor expansion: $\omega^2 = c_0^2k^2 - \frac{1}{3}c_0^2h_0^2k^4$.

The above correction of the Saint-Venant model generates the following form of what are called the Boussinesq equations

$$\frac{\partial h}{\partial t} + \frac{\partial (vh)}{\partial x} = 0$$

$$\frac{\partial v}{\partial t} + v\frac{\partial v}{\partial x} + g\frac{\partial h}{\partial x} + \frac{1}{3}c_0^2h_0^2\frac{\partial^3 h}{\partial x^3} = 0\,,$$

that describes water waves that combine dispersive effects of the order of $(h_0/l)^2$ (where l is the wavelength) and nonlinear effects of the order of a/h_0 (where a is the wave amplitude).

The Boussinesq equations describe waves that propagate in both directions along x-axes. If waves moving only to the right direction are considered and the bulk velocity v is eliminated, the KdV model is derived.

□

Example 1.2.2.b
Generalizations of the Korteweg–de Vries Model

Several generalizations of the KdV model have been proposed: for instance, a generalized (dimensionless) third-order KdV model is written as

$$\frac{\partial u}{\partial t} + u^m \frac{\partial u}{\partial x} + \mu \frac{\partial^3 u}{\partial x^3} = 0 \cdot \qquad (1.2.10)$$

Applications of this model take advantage of some analytical solutions which can be compared with computational ones (see Chapter 3 for details).

A further generalization of the original KdV model (1.2.8) is the following fifth-order KdV equation, called the Kawahara equation (see Parkes and Duffy (1996)):

$$\frac{\partial u}{\partial t} + u\frac{\partial u}{\partial x} + \frac{\partial^3 u}{\partial x^3} - \frac{\partial^5 u}{\partial x^5} = 0. \qquad (1.2.11)$$

The presence of the fifth-order derivative makes this model an appropriate candidate to test numerical simulations.

Finally, the KdV model can also be modified introducing variable coefficients, as proposed by Johnson (1997):

$$2\sqrt{D(X)}\,\frac{\partial u}{\partial x} + \frac{1}{2}\frac{D'(X)}{\sqrt{D(X)}}\,u + \frac{3}{D(X)}\,u\frac{\partial u}{\partial \xi} + \frac{1}{3D(X)}\frac{\partial^3 u}{\partial \xi^3} = 0\,, \qquad (1.2.12)$$

with

$$\xi = \frac{1}{\epsilon}\chi(X) - t\,, \quad X = \epsilon x\,, \quad \chi(X) = \int_0^X \frac{dX'}{\sqrt{D(X')}}\,, \qquad (1.2.13)$$

where $D(X)$ represents the local depth while ϵ stands for the maximum amplitude/depth. This model describes a nonlinear dispersive wave that propagates over a variable depth.

□

Example 1.2.3
The Sine-Gordon Model

The sine-Gordon model, see Whitham (1974), in its dimensionless form is written as

$$\frac{\partial^2 u}{\partial t^2} - \frac{\partial^2 u}{\partial x^2} + \sin u = 0\,. \qquad (1.2.14)$$

This is a particular form of the Klein–Gordon equation

$$\frac{\partial^2 u}{\partial t^2} - \frac{\partial^2 u}{\partial x^2} + v'(u) = 0\,,$$

where v' is a nonlinear function of u regarded as the derivative, with respect to u, of a potential energy $V(u)$. Model (1.2.14) is the most widely studied in the literature as it refers to various interesting physical systems where the source term shows an oscillatory behavior related to the state variable.

It was formulated for the first time to study the propagation of a slip in a one-dimensional crystal to investigate the geometry of pseudo-spherical surfaces (i.e., surfaces of constant negative curvature, Whitham (1974), Shen (1993)). Since then the sine-Gordon equation has been used to model several phenomena as documented in the review by Dodd et al. (1982). Among others, mention can be made of the Josephson junction phenomenon in transmission lines, where $\sin u$ is the current across an insulator between two superconductors and the voltage is proportional to $\partial_t u$; of the dislocation in one-dimensional solid crystals, where u describes the crystal displacement; of the propagation in ferromagnetic or anti-ferromagnetic fields, where u is the angle between the direction of magnetization and that of the external magnetic field, etc.

The sine-Gordon model exhibits a travelling wave solution and both periodic and solitary solutions (Whitham (1974). In particular, the solution for two interacting solitary waves (e.g., Shen (1993)) is written as

$$u(x,t) = 4\arctan\left[\frac{\sqrt{m^2-1}}{m}\,\frac{\sinh(mx+c_1)}{\cosh(\sqrt{m^2-1}t+c_2)}\right], \quad m>1, \qquad (1.2.15)$$

where m, c_1, and c_2 are integration constants.

Another analytical solution useful to test numerical procedures is called the breather solution (Shen (1993))

$$u(x,t) = 4\arctan\left[\frac{m}{\sqrt{1-m^2}}\,\frac{\sin(\sqrt{1-m^2}t+c_2)}{\cosh(mx+c_1)}\right], \quad -1<m<1. \qquad (1.2.16)$$

□

Example 1.2.4
River Pollution Model

The modelling of the fate of chemicals in rivers is an important topic in environmental studies (e.g., Fischer et al. (1979) Schnoor (1996)). We refer to a one-dimensional convection-dispersion model with spatially variable coefficients and nonlinear decay involving a source term.

Let us consider a steady, turbulent open channel flow with a mild slope i_b. The depth h and the mean velocity v of the stream vary along the channel according to the following power laws:

$$h = a_h \left(1 + \frac{\xi}{\ell}\right)^{b_h} \quad \text{and} \quad v = a_v \left(1 + \frac{\xi}{\ell}\right)^{b_v}, \tag{1.2.17}$$

where ξ is the longitudinal coordinate, a_h and a_v are dimensional constants, b_h and b_v are dimensionless constants, and ℓ is a reference length, which will be assumed to be equal to the length of the channel. The previous relationships are commonly used to model the changes of the hydraulic characteristics along rivers, where the coefficients and exponents are suitable experimental parameters that depend on the type of river. See, e.g., Leopold and Maddock (1953) and Gupta and Cvetkovic (2000).

The advection and dispersion of a chemical in the stream is governed (Fischer et al. (1979)) by the following model:

$$\frac{\partial C}{\partial \tau} = -\frac{\partial}{\partial \xi}\left(v(\xi)C\right) + \frac{\partial}{\partial \xi}\left(K(\xi)\frac{\partial C}{\partial \xi}\right), \tag{1.2.18}$$

where $C = C(\xi, \tau)$ is the depth-averaged concentration of the chemical per volume of water, τ is the time, $v(\xi)$ is the transport velocity defined in Eq. (1.2.17), and $K = K(\xi)$ is the dispersion coefficient.

The latter coefficient can be modelled using the results by Elder (1959): in a channel flow, dispersion can be described by the phenomenological model

$$K = K(\xi) = k_0\, u_\star(\xi)\, h(\xi), \qquad k_0 = 5.93\,, \tag{1.2.19}$$

where $u_\star$ is the friction velocity of the stream.

Therefore, recalling that

$$u_\star(\xi) = \sqrt{g\, i_b\, h(\xi)}\,, \tag{1.2.20}$$

where g is the gravity, and the dispersion coefficient h depends on ξ, according to the power law

$$K = a_K \left(1 + \frac{\xi}{\ell}\right)^{b_K}, \tag{1.2.21}$$

where a_K is a dimensional constant and b_K is a dimensionless constant that is equal to

$$a_K = k_0 \sqrt{g i_b}\, a_h^{3/2} \qquad \text{and} \qquad b_K = \frac{3}{2} b_h. \tag{1.2.22}$$

Generalizations of the model (1.2.18) can be obtained by adding suitable source terms and including nonlinear decay reactions. The decay is modelled according to the power law (Schnoor (1996)): $R = -\lambda C^m$, where λ is the decay rate constant and $m > 1$ in order to model the common chemodynamic nonlinearities. The source term is assumed to be of the shape

$$cS(\tau)F(\xi)\,, \tag{1.2.23}$$

where c is a dimensional constant, and $S = S(\tau)$ and $F = F(\xi)$ are smooth dimensionless functions of time and space, respectively.

In this case, the model (1.2.18) is written as

$$\begin{aligned}\frac{\partial C}{\partial \tau} &= -\frac{\partial}{\partial \xi}(v(\xi)C) + \frac{\partial}{\partial \xi}\left(K(\xi)\frac{\partial C}{\partial \xi}\right) - \lambda C^m + cS(\tau)F(\xi)\\ &= \left(\frac{\partial K(\xi)}{\partial \xi} - v(\xi)\right)\frac{\partial C}{\partial \xi} + K(\xi)\frac{\partial^2 C}{\partial \xi^2}\\ &\quad - \left(C\frac{\partial v(\xi)}{\partial \xi} + \lambda C^m\right) + cS(\tau)F(\xi)\,,\end{aligned} \tag{1.2.24}$$

where $\xi \in [0, \ell]$, $\tau \in [0, T]$, and $C \in [0, C_{max}]$, and where $v(\xi)$, $K(\xi)$, $F(\xi)$, and $S(\tau)$ have been defined above.

□

Example 1.2.5
First-Order Models of Vehicular Traffic Flow

Hydrodynamic modelling of a one-lane flow of vehicles on a road can be obtained, as shown in the review paper by Bellomo, Delitala, and Coscia (2002), using the mass conservation equation already introduced for Example 1.2.1. In fact, the derivation follows precisely the same line.

The modelling refers to the following dimensionless variables: $t = t_r/T$, that is, the time variable that refers to the characteristic time T, where t_r is the real time; and $x = x_r/\ell$, that is, the space variable that refers to the characteristic length of the road ℓ, where x_r is the real dimensional space; both x and t are independent variables.

The first dependent variable $u = n/n_M$ is the density that refers to the maximum density n_M of vehicles in a bumper-to-bumper traffic jam, while the second dependent variable $v = v_R/v_M$ is the velocity that refers to the maximum mean velocity v_M, where v_R is the real velocity of vehicles; $q = uv$ is the linear mean flux that refers to the maximum admissible mean flux $q_M = n_M v_M$. The characteristic time T is given by $v_M T = \ell$, which means that T is the time necessary to cover the whole length of the road at the maximum mean velocity.

The mass conservation equation related to the variables $u = u(t,x) \in [0,1]$ and $v = v(t,x) \in [0,1]$ is written as

$$\frac{\partial u}{\partial t} + \frac{\partial}{\partial x}(uv) = 0 \cdot \tag{1.2.25}$$

The modelling takes advantage of the experimental information delivered by the analysis of steady uniform flow conditions. The phenomenological behavior of the system shows that the mean velocity of the car decays with increasing density from the value $v = 1$ when $u \cong 0$ to $v = 0$ when $u = 1$. The simplest way of representing this behavior is the following:

$$v_e = v_e(u) = 1 - u\,, \qquad q_e = uv_e = u(1-u)\,, \tag{1.2.26}$$

while a relatively more precise interpretation of experiments related to the ***velocity diagram*** was proposed in the paper by Bonzani and Mussone (2003) by the following interpretation of experimental data:

$$v = \exp\left\{ -\alpha \frac{u}{1-u} \right\}, \qquad \alpha > 0\,.$$

The closure of Eq. (1.2.25) can be obtained by using (1.2.26) and replacing u by the ***apparent local density*** $u^\star$ which is related to the fact that the driver is not able to measure the local density exactly, but simply feels it. In other words, the driver feels a density $u^\star$ that is larger than the real one if the local density gradient is positive, and smaller than the real one if the gradient is negative. In addition, the above multiplicative effect increases with decreasing density.

According to these phenomenological considerations, the following expression is proposed for the local fictitious density:

$$u^\star = u\left[1 + \eta(1-u)\frac{\partial u}{\partial x}\right], \tag{1.2.27}$$

where η is a positive parameter. Technical calculations then yield

$$\frac{\partial u}{\partial t} + (1-2u)\frac{\partial u}{\partial x} = \eta u^2(1-u)\frac{\partial^2 u}{\partial x^2} + \eta u(2-3u)\left(\frac{\partial u}{\partial x}\right)^2 . \tag{1.2.28}$$

The same model, using the flow as an additional dependent variable, can be rewritten as a system of two equations:

$$\begin{cases} \dfrac{\partial u}{\partial t} = -\dfrac{\partial q}{\partial x}\,, \\[2mm] \dfrac{\partial q}{\partial t} = \left[(1-2u) + \eta u(2-3u)\dfrac{\partial q}{\partial x}\right]\dfrac{\partial q}{\partial x} + \eta u(1-u)\dfrac{\partial^2 q}{\partial x^2}\,. \end{cases} \tag{1.2.29}$$

The same model can be revisited using the phenomenological model proposed by Bonzani and Mussone (2003) to relate the mass velocity to the local density. However, for tutorial purposes the relatively simpler model (1.2.28) appears to be suitable for the application of the method proposed in Chapter 4.

□

Example 1.2.6
Population Dynamics with Internal Structure

This example refers to the population dynamics model with kinetic interaction, which was proposed by Jager and Segel (1992) to study the evolution of a physical state, called ***dominance***, that characterizes certain populations of insects. The model consists of an evolution problem, in terms of a nonlinear system of integro-differential equations, that define, for each population, the evolution of the probability density function over the dominance.

The physical system is constituted by n interacting populations. Each population individual can be found in a state described by a variable $u \in [0, 1]$, which can be called the ***microscopic state***, and which should be regarded as a dimensionless, normalized, real independent variable. The description of the overall state of the system is given by the probability density function that is defined, for each population, by

$$f_i = f_i(t, u) \, : \, [0, T] \times [0, 1] \to \mathbb{R}_+ \,. \tag{1.2.30}$$

Therefore, the probability of finding, at time t, an individual of the ith population in the state interval $[u_1, u_2]$ is given by

$$\mathcal{P}_i(t, u \in [u_1, u_2]) = \int_{u_1}^{u_2} f_i(t, u)\, du \,. \tag{1.2.31}$$

The mathematical model is derived assuming that the microscopic state is modified by encounters and that only binary encounters play a role in the game. Moreover, the derivation needs a detailed analysis of microscopic interactions, which can be identified by the following quantities:

• The encounter rate between pairs of individuals, of the ith population in the state u and of the jth population in the state v, is identified by the term $\eta_{ij}(u,v) \geq 0$.

• The probability that an individual of the ith population in the state v ends up in the state u conditionally to an encounter with an individual of the jth population in the state w has a density with respect to the variable u denoted by $\varphi_{ij}(v,w;u) \geq 0$.

If no external action modifies the internal state, the expression of the mathematical model follows:

$$\begin{aligned}\frac{\partial f_i}{\partial t}(t,u) &= J_i[f](t,u)\\ &= \sum_{j=1}^{n}\left\{\int_0^1\int_0^1 \eta_{ij}(v,w)\varphi_{ij}(v,w;u)f_i(t,v)f_j(t,w)\,dv\,dw\right.\\ &\quad\left. - f_i(t,u)\int_0^1 \eta_{ij}(u,v)f_j(t,v)\,dv\right\}. \end{aligned} \tag{1.2.32}$$

The assumption that φ is a probability density is written as

$$I = \int_0^1 \varphi_{ij}(v,w;u)\,du = 1\,, \quad \forall\, i,j\,,v,w\,, \tag{1.2.33}$$

which means that the number of individuals of each population is preserved. Otherwise, if $I > 1$ (respectively $I < 1$), the number of individuals increases (or respectively decreases).

□

The above model is valid in the spatially homogeneous case under the further assumption that the variable u is not subject to external actions. The following generalizations can be proposed to describe the above-described phenomena.

Example 1.2.7
Generalized Population Dynamics

Various generalizations of model (1.2.32) have been proposed by Arlotti et al. (2000). Let us here consider the model described in Example 1.2.6 under the assumption that the variable u is subject to an external action

$k(u)$. The related flow due to transport is therefore $k(u)f$ so that the model, in the case of one population only, is written as

$$\frac{\partial f}{\partial t}(t,u) + \frac{\partial}{\partial u}(k(u)f(t,u)) = J[f](t,u)\,, \tag{1.2.34}$$

where

$$J[f](t,u) = \left\{ \int_0^1 \int_0^1 \eta(v,w)\varphi(v,w;u)\, f(t,v)f(t,w)\, dv\, dw \right.$$
$$\left. - f(t,u) \int_0^1 \eta(u,v) f(t,v)\, dv \right\}. \tag{1.2.35}$$

An additional generalization refers to the case of models with a space structure. In this case, a diffusion term is added to the left-hand side of the equation. A typical example is as follows, in a one space dimension:

$$\frac{\partial f}{\partial t}(t,u,x) + \frac{\partial}{\partial u}(k(u)f(t,u,x)) = J[f](t,u,x) + \varepsilon\,\frac{\partial^2 f}{\partial x^2}(t,u,x)\,. \tag{1.2.36}$$

□

Example 1.2.8
A Spatial Reaction-Diffusion Model

Spatial reaction-diffusion models represent one of the most investigated forms of partial differential equations; see, e.g., Grindrod (1996) Hundsdorfer and Verwer (2003). They have attracted a lot of attention in particular because of their possibility of generating spatial patterns, and this explains the widespread use of reaction-diffusion models in several scientific fields, e.g., biology, chemistry, and fluid dynamics. See, e.g., Cross and Hohenberg (1993), Murray (1993).

We here consider a model introduced by Schnakenberg (1979) to describe a series of trimolecular autocatalytic reactions. The model consists of the following system of reaction-diffusion equations in two space dimensions:

$$\begin{cases} \dfrac{\partial u}{\partial t} = -\kappa(a - u + u^2 v) + \varepsilon\left(\dfrac{\partial^2 u}{\partial x^2} + \dfrac{\partial^2 u}{\partial y^2}\right), \\[2ex] \dfrac{\partial v}{\partial t} = \kappa(b - u^2 v) + \left(\dfrac{\partial^2 v}{\partial x^2} + \dfrac{\partial^2 v}{\partial y^2}\right). \end{cases} \tag{1.2.37}$$

where u and v are dimensionless concentrations of self-activating and self-inhibiting chemical substances, respectively; and κ, a, b, and ε are dimensionless parameters.

The terms $\kappa\, a$, $\kappa\, u$, and $\kappa\, u^2\, v$ model, in the first differential equation, the decay, self-activation, and coupling with the second equation, respectively. The last term on the right-hand side of the first equation models spatial diffusion phenomena. Similarly, the first, second, and third terms in the second equation model growing phenomena, interaction with the first variable, and diffusion, respectively.

The interesting aspect of this model is its capability to select peculiar spatial patterns where initial small perturbations are amplified and spread, leading to the formation of spots that slowly move and interact. If relatively large diffusion coefficients are used, the model is stiff and, for this reason, the Schnakenberg model is a good candidate to test numerical algorithms (e.g., Hundsdorfer and Verwer (2003)).

□

1.3 Dimensional Analysis

Some of the mathematical models described in Section 1.2 have already been written in terms of dimensionless variables. This means that both the independent and dependent variables are divided by suitable reference variables related to the physical system which is the subject of the modelling process. This is the case, for instance, in Example 1.2.5 where we described the characteristic density lengths and time in detail. This feature needs to be technically generalized considering that the application of mathematical methods can take advantage of the above ***scaling*** (or ***dimensional***) analysis, see Barenblatt (2003).

The scaling technique consists in the partial (or even the full) removal of physical units from a given mathematical equation by pertinent substitution of variables. From a physical viewpoint, such a scaling is based on the fact that some of the quantities involved in the model can be handled in a more adequate way if they are measured relative to some appropriate units. These units refer to quantities *intrinsic to the system*, in contrast to the usual units one deals with (such as SI units). For a system of equations modelling a given physical process, the scaling procedure generally follows four steps: the first one is of course to identify all the independent and dependent variables of the system. Then, one proceeds in the substitution of each of them with a quantity scaled relative to some characteristic unit of measure to be determined. Then a more delicate step consists in the appropriate choice of the characteristic unit for each variable so that the coefficients of as many terms as possible become equal to one. The final step consists of writing the system of equations in terms of their new dimensionless variables.

Let us illustrate this procedure with an example of general first models stated in terms of partial differential equations in a bounded two-dimensional domain and let $u_r = u_r(t_r, x_r, y_r)$ be the dimensional dependent variable which describes the state of the system related to the dimensional independent variables t_r, x_r, and y_r. Then, all variables can be referred to the minimum and maximum values of each variable so that each one is defined over the interval $[0, 1]$. The solution of problems may, in some cases, give a result outside the above interval. However, this is not the point as it is sufficient to have variables of the order of one.

Considering the dependent variable, with obvious meanings of symbols, the following dimensionless dependent variable is obtained:

$$u = \frac{u_r - u_m}{u_M - u_m} \,. \tag{1.3.1}$$

The same reasoning can be applied to the space variables defined over a rectangle $[x_m, x_M] \times [y_m, y_M]$:

$$x = \frac{x_r - x_m}{x_M - x_m}, \qquad \text{and} \qquad y = \frac{y_r - y_m}{x_M - x_m},$$

where $x_M - x_m$ is chosen as a characteristic length for the space variables. The dependent variable $u = u(t, x, y)$ defines an application from $[0, 1] \times [0, 1] \times [0, \ell]$ into $[0, 1]$ where

$$\ell = \frac{y_M - y_m}{x_M - x_m} \,.$$

Analogous normalizations can be made when the space variables are defined over nonrectangular domains. On the other hand, when problems are such that the dependent variables are defined over an unbounded domain, the variable can refer to any length that has a physical meaning, while it can only refer to the interval $[0, 1]$ after a suitable change in variables.

The choice of the reference time, say T_c, is technically more complex. It must, in fact, be related to the actual analytic structure of the model, while trying to bring the cause and effect identified in the model to the same order. In most of the models of Section 1.2, the ***cause*** is identified by the right-hand side term, while the ***effect*** is the left-hand term. This is precisely what has been done in the case of the traffic flow model. The examples which follow should further clarify the above topic.

Example 1.3.1
Linear and Nonlinear Diffusion Models

Let us consider the diffusion model described in Example 1.2.1 obtained when $c = 0$ and $K = K_0$ is a constant. The model, in terms of real variables, can be written as follows:

$$\frac{\partial c_r}{\partial t_r} = K_0 \frac{\partial^2 c_r}{\partial x_r^2}. \tag{1.3.2}$$

Assuming $u = c_r/c_M$, $t = t_r/T_c$, and $x = x_r/\ell$, yields

$$\frac{1}{T_c}\frac{\partial u}{\partial t} = \frac{K_0}{\ell^2}\frac{\partial^2 u}{\partial x^2}. \tag{1.3.3}$$

Moreover, taking

$$\frac{K_0 T_c}{\ell^2} = 1 \quad \text{or equivalently} \quad T_c = \frac{\ell^2}{K_0},$$

provides the following expression of the model:

$$\frac{\partial u}{\partial t} = \frac{\partial^2 u}{\partial x^2}. \tag{1.3.4}$$

The time evolution is then analyzed in terms of units of T_c.

The same reasoning can be applied to the nonlinear diffusion model where the diffusion coefficient is given by $k_0 K(u)$. The following dimensionless model is then obtained:

$$\frac{\partial u}{\partial t} = K_u(u)\left(\frac{\partial u}{\partial x}\right)^2 + K(u)\frac{\partial^2 u}{\partial x^2}, \tag{1.3.5}$$

where $K_u(\cdot)$ denotes the partial derivative of K with respect to u.

□

The reasoning applied to the above simple application can be technically followed to deal with relatively more complex models such as the one reported in Example 1.2.5. It is interesting to note how this method generates dimensionless numbers which have a well-defined physical meaning. This peculiarity is shown in the example that follows.

Example 1.3.2
Dimensionless River Pollution

Again considering Example 1.2.4, let us define the following dimensionless variables: $u = C/C_M, x = \xi/\ell$, and $t = \tau/T_c$, where T_c is a critical time selected in order to yield the coefficient of the term with higher-order spatial derivative lower or equal to one. Technical calculations yield

$$\frac{T_c a_K}{\ell^2} = \varepsilon \leq 1, \quad \text{i.e.,} \quad T_c = \varepsilon \frac{\ell^2}{a_K},$$

where ε should be chosen with the aim of keeping T_c consistent with the time necessary to observe the phenomena.

If the previous dimensionless variables are introduced into the model (1.2.24), the following dimensionless equation is obtained:

$$\begin{aligned} \frac{\partial u}{\partial t} &= [\varepsilon_1 f_1(x) - \varepsilon_2 f_2(x)] \frac{\partial u}{\partial x} + \varepsilon_3 f_3(x) \frac{\partial^2 u}{\partial x^2} \\ &\quad - \mu u^m - \varepsilon_4 f_4(x)\, u + \eta\, s(t)\, q(x)\,, \end{aligned} \tag{1.3.6}$$

where the dimensionless parameters $(\varepsilon_1 - \varepsilon_4)$ are defined as follows:

$$\varepsilon_1 = \frac{a_K b_K T_c}{\ell^2}, \quad \varepsilon_2 = \frac{a_v T_c}{\ell}, \quad \varepsilon_3 = \frac{a_K T_c}{\ell^2} = \frac{\varepsilon_1}{b_K},$$

$$\varepsilon_4 = \frac{a_v b_v T_c}{\ell} = \varepsilon_2 b_v\,,$$

while the functions of the space variable are

$$f_1(x) = (1+x)^{b_K - 1}, \qquad f_2(x) = (1+x)^{b_v},$$

$$f_3(x) = (1+x)^{b_K}, \qquad f_4(x) = (1+x)^{b_v - 1},$$

and the constants μ and η are $\mu = \lambda C_M^{m-1} T_c$ and $\eta = c T_c C_M^{-1}$.

□

Example 1.3.3
Dimensionless Population Dynamics

Let us consider the population dynamics model introduced in Example 1.2.6. The microscopic state variable $u \in [0, 1]$ is a normalized dependent variable obtained from a real state variable u_r through

$$u = \frac{u_r - a}{b - a},$$

where the real state variable u_r is in the real interval $I = [a, b] \subset \mathbb{R}$. The densities f_i, for models with a constant number of individuals, can be normalized in such a way that their integration over u is equal to one: the real distribution is divided, for each population, by the number of individuals of the population. If the model includes the space variable x, then this independent variable should also be put in a suitable dimensionless form.

□

1.4 Classification of Models and Mathematical Problems

As already mentioned, the application of the mathematical collocation method refers to mathematical problems that are obtained by linking the model (for instance a partial differential equation) to the ***initial conditions***, namely the value of the dependent variable for $t = 0$, and to the ***boundary conditions***, namely the values on the boundary of the domain of the independent variable.

As known, the statement of the problems needs to follow a certain number of rules which are necessary to assume existence and uniqueness of solutions as well as their regular dependence on the data of the problem. It is useful to provide a classification of the mathematical equations (or models) to obtain a correct statement of problems. Such a classification also contributes to a better understanding of some qualitative properties of the solutions.

Ordinary differential equations are usually classified according to the order of the higher derivatives, but the classification of partial differential equations is a more delicate matter. Several methods of classification coexist, some of them based upon Fourier analysis, others upon spectral analysis. The classification summarized in what follows should be regarded as a concise introduction to a topic that certainly needs additional analysis, and which can be developed with reference to specialized literature, e.g., Evans (1998). The analysis is first related to second-order partial differential equations, and a relatively simple analysis of first-order equations is then proposed.

Let us consider first the following class of second-order partial differential equations in two independent variables:

$$A(x)\frac{\partial^2 u}{\partial t^2} + 2B(x)\frac{\partial^2 u}{\partial t \partial x} + C(x)\frac{\partial^2 u}{\partial x^2} = F\left(t, x; u, \frac{\partial u}{\partial t}, \frac{\partial u}{\partial x}\right). \qquad (1.4.1)$$

Some important mathematical properties of the solution are essentially determined by the left–hand side term, which contains the highest-order derivatives and the classification of such equations referred only to the coefficient of higher-order derivatives. Just as one classifies *conic sections* as parabolic, hyperbolic, and elliptic based on the *discriminant*, the same can be done for the above second-order partial differential equation *at a given point*:

– If $B^2 - AC > 0$ then the equation is said to be ***hyperbolic***.

– If $B^2 - AC < 0$ then the equation is ***elliptic***.

– If $B^2 - AC = 0$, then the equation is ***parabolic***.

Of course, the sign of the discriminant $B^2 - AC$ can also depend on space through the local values of the coefficients A, B, and C. Therefore, the classification can change in space.

It turns out that such a classification corresponds to real values, null, and imaginary values, respectively, of the eigenvalues of the symmetric matrix with diagonal entries $A(x)$ and $C(x)$ and off-diagonal entry $B(x)$.

The above classification can be related to the following class of second-order equations in several space variables:

$$\sum_{i,j=1}^{n} A_{ij}(\mathbf{x}) \frac{\partial^2 u}{\partial x_i \partial x_j} = F\left(\mathbf{x}; u, \frac{\partial u}{\partial x_1}, \dots, \frac{\partial u}{\partial x_n}\right), \tag{1.4.2}$$

where $\mathbf{A} = (A_{ij}(\mathbf{x}))_{ij}$ is a symmetric matrix (note, in particular, that its eigenvalues are real). Then, it is possible to classify the types of the above equation with respect to the sign of the eigenvalues of $\mathbf{A}$:

– If $\mathbf{A}$ admits (at least) a *zero* eigenvalue (i.e., $\det \mathbf{A} = 0$), then the equation is ***parabolic***.

– If all the eigenvalues of $\mathbf{A}$ are nonzero and of the same sign, then the equation is ***elliptic***.

– If all the eigenvalues of $\mathbf{A}$ are nonzero and all *all but one* have the same sign, then the equation is ***hyperbolic***.

The above classification is not exhaustive. In particular, the case of a matrix $\mathbf{A}$ with nonzero eigenvalues but with several of them changing signs does not enter our classification.

Clearly, since the coefficients of $\mathbf{A}$ are not constant, its eigenvalues may depend on the space variable x through the local values of the coefficients $A_{ij}(\mathbf{x})$. In such a case, the associated equation may not belong to any of these categories but rather be of ***mixed type***. A simple but important example is the Euler–Tricomi equation:

$$u_{xx} = x u_{yy}$$

which is called ***elliptic-hyperbolic*** since it is elliptic in the region $\{x < 0\}$, hyperbolic in the region $\{x > 0\}$, and (degenerate) parabolic on the line

$x = 0$. The numerical investigation of such mixed-type equations is far more delicate.

A similar classification can also be proposed for general first-order equations of the type:

$$\mathcal{A}\frac{\partial \mathbf{u}}{\partial t} + \mathcal{B}\frac{\partial \mathbf{u}}{\partial x} = \mathbf{f}\,, \tag{1.4.3}$$

where $\mathcal{A} = (A_{ij})_{i,j}$, $\mathcal{B} = (B_{ij})_{i,j}$, and $\mathbf{f} = \{f_i\}_i$ can depend on the space variable; $\mathbf{f}$ can also depend on the dependent variable $\mathbf{u} = (u_1, \ldots, u_n)$.

Under the assumption that $\mathcal{A}$ is nonsingular (i.e., $\det \mathcal{A} \neq 0$), the classification is based on the roots of the characteristic polynomial:

$$P_n(\lambda) = \det(\mathcal{B} - \lambda\mathcal{A}) = 0\,, \tag{1.4.4}$$

as well as on the number k of the independent eigenvectors $\mathbf{v}_i$ $(i = 1, \ldots, k)$ associated to the eigenvalue problem:

$$(\mathcal{B} - \lambda\mathcal{A})^\top \mathbf{v}_i = \mathbf{0}\,. \tag{1.4.5}$$

Precisely, the characteristic polynomial $P_n(\lambda)$ has degree n, and the classification is related to the reality of its root and the number of linearly independent solutions to (1.4.5):

– The system is ***parabolic*** when $P_n(\lambda)$ has n real zeros at least one of which is repeated and $k < n$;

– The system is ***elliptic*** if $P_n(\lambda)$ has no real zeros.

– Finally, it is ***hyperbolic*** if $P_n(\lambda)$ has n real distinct zeros, or if $P_n(\lambda)$ has n real zeros at least one of which is repeated and $k = n$.

The above classification is useful for understanding the qualitative difference of the solution to mathematical problems corresponding to elliptic, hyperbolic, and parabolic models. Some preliminary indications can be given to cover some introductory aspects of the general problem, which certainly deserves a deeper analysis.

• ***Hyperbolic equations*** are evolution equations that describe wave-like phenomena. The solution of a hyperbolic problem in unbounded domains cannot be smoother than the initial data. It can actually develop singularities, as time goes by, even from smooth initial data, which then characterize the whole evolution and are propagated along special curves called ***characteristics***. If the solution initially has a compact support, then the rate at which the support expands can be interpreted as the ***speed of propagation***.

• ***Parabolic equations*** are equations that describe diffusion-like phenomena and are typical for dissipative processes. In general, the solution of a parabolic problem is infinitely differentiable with respect to both space and time even if the initial data are not continuous. In other words, singularities can neither develop nor be maintained (***smoothing effect***). Moreover, even if the initial data have compact support, perturbations are felt everywhere. This is usually indicated by saying that parabolic systems have an ***infinite speed of propagation***.

• ***Elliptic equations*** describe systems in an equilibrium or steady state. They can be seen as equations that describe, for instance, the final state reached by a physical system described by a parabolic equation after the transient term has died out. Elliptic equations are also well suited for physical problems related to the determination of potential problems. Solutions of elliptic partial differential equations are as smooth as the coefficients allow, within the interior of the region where the equation and solutions are defined. For example, solutions to Laplace equation are analytic within the domain where they are defined, but solutions may assume boundary values that are not smooth. As is the case for parabolic problems, any local perturbation influences the whole domain even if the magnitude of such an influence is decreasing with increasing distance from the (localized) source of the perturbation. Another peculiar feature is the applicability of ***maximum principles*** which assert that the maximum of the solution to some elliptic equation is reached only at the boundary.

The preceding considerations should be considered as very preliminary information which may give engineers and applied scientists a rough idea of the qualitative behavior of solutions in view of the application of computational algorithms. More detailed information can be found in specialized literature, e.g., Dautray and Lions (1990). The reader can develop some practice on the various models in this section; models of all the above types have been proposed.

Moreover, models that include both first- and second-order terms may be hyperbolic with respect to the first-order term and parabolic with respect to the second-order term. This is, in fact, the case for the nonlinear diffusion model that corresponds to Example 1.2.1 and the traffic flow model that corresponds to Example 1.2.5. These models become elliptic when the time derivative is placed equal to zero.

The application of models to the analysis of real-world systems generates mathematical problems that are obtained by adding the conditions necessary to find quantitative solutions to the evolution equation. These conditions refer to the behavior of the dependent variable at the boundaries of the dependent variables domain.

Let us first consider scalar models that involve the whole set of space variables, say $\mathbf{x} = \{x_1, x_2, \dots, x_n\}$, with $(n \geq 1)$; the boundary conditions have, consequently, to be assigned for all the space variables. Let us, in particular, consider the case of models that can be written as follows:

$$\frac{\partial u}{\partial t} = f\left(t, \mathbf{x}, \frac{\partial u}{\partial x_1}, \ldots, \frac{\partial u}{\partial x_n}, \frac{\partial^2 u}{\partial x_1^2}, \ldots, \frac{\partial^2 u}{\partial x_n^2}\right), \tag{1.4.6}$$

where u is the dependent variable

$$u = u(t, \mathbf{x}) : \quad [0, 1] \times D \to \mathbb{R}, \tag{1.4.7}$$

and where the boundary of the $D \subseteq \mathbb{R}^n$ domain of the space variables is denoted by ∂D. Some general definitions and rules can be given.

• The ***initial conditions*** are defined by the values of u at $t = 0$ for all $\mathbf{x} \in D$: $u_0(\mathbf{x}) = u(t = 0, \mathbf{x})$.

• The ***boundary conditions*** are defined by the values, for all $t \geq 0$ and $\mathbf{x}$ at the boundary ∂D of D, of u or its normal derivative with respect to ∂D:

$$u_d(t, \mathbf{x}) = u(t, \mathbf{x} \in \partial D), \tag{1.4.8}$$

or

$$u_{\mathbf{n}}(t, \mathbf{x}) = \frac{\partial u}{\partial \mathbf{n}}(t, \mathbf{x} \in \partial D) = \mathbf{n}(\mathbf{x}) \cdot \nabla u(t, \mathbf{x}), \tag{1.4.9}$$

where $\mathbf{n}$ is the normal (when it is well defined) to ∂D directed inside D.

• The ***statement of the problem*** is defined by assigning, if both time and space derivatives appear in the mathematical model, both initial and boundary conditions. The relative mathematical problem is called an ***initial-boundary value problem.*** If, instead, the mathematical model is static, i.e., time independent, then, of course, no initial conditions are needed. In this case, the mathematical problem is called a ***boundary value problem.*** If the model is constituted by a system of equations, then initial and/or boundary conditions have to be assigned for all equations.

The above general rules can be specialized to particular problems. Let us again consider the class of models defined by Eq. (1.4.6). The following problems can be stated.

Problem (Dirichlet) 1.4.1. *The initial-boundary value problem for the scalar Eq. (1.4.6) with Dirichlet boundary conditions is stated with initial condition*

$$u(0, \mathbf{x}) = u_0(\mathbf{x}), \qquad \forall \mathbf{x} \in D, \tag{1.4.10}$$

and boundary conditions

$$\forall t \in [0, 1], \quad \forall \mathbf{x} \in \partial D : \quad u(t, \mathbf{x} \in \partial D) = u_d(t, \mathbf{x}), \tag{1.4.11}$$

given as functions that are consistent, for $t = 0$, with the initial condition (1.4.10).

Problem (Neumann) 1.4.2. *The initial-boundary value problem for the scalar Eq. (1.4.6) with Neumann boundary conditions is stated with the initial condition (1.4.10) and boundary conditions*

$$\forall t \in [0,1], \quad \forall \mathbf{x} \in \partial D : \quad \frac{\partial u}{\partial \mathbf{n}}(t, \mathbf{x} \in \partial D) = u_{\mathbf{n}}(t, \mathbf{x}), \tag{1.4.12}$$

given as functions of time and space and where $\mathbf{n}$ *denotes the normal to* ∂D *directed inside* D.

Relatively more general cases can be technically developed starting from the indications given for the above classes of models. For instance, one can complement Eq. (1.4.6) by the Neumann boundary conditions on some part $\partial D_{\mathbf{n}}$ of ∂D and by the Dirichlet boundary conditions on another part ∂D_d, where $\partial D = \partial D_d \cup \partial D_{\mathbf{n}}$ (***mixed boundary conditions***).

The value of some linear combination of u and its normal derivative on ∂D (***Robin boundary conditions***) can also be prescribed.

Finally, we mention that ***nonlinear boundary conditions*** can also be envisaged, that prescribe a nonlinear combination of u and its normal derivative:

$$\forall t \in [0,1], \quad \forall \mathbf{x} \in \partial D : g(t, \mathbf{x}, u(t, \mathbf{x}), \frac{\partial u}{\partial \mathbf{n}}(t, \mathbf{x})) = 0, \tag{1.4.13}$$

where g is a suitable function of its arguments.

Remark 1.4.1. *The above statements can be particularized in the case of models in one space dimension. In fact, if* $x \in D = [0,1]$, *and since* ∂D *is only made of two points* $\partial D = \{0\} \cup \{1\}$, *it is possible to recast, e.g., the Neumann boundary conditions as*

$$\frac{\partial u}{\partial x}(t, 0) = \gamma(t), \quad \text{and} \quad \frac{\partial u}{\partial x}(t, 1) = \delta(t), \qquad \forall t \in [0,1], \tag{1.4.14}$$

where γ *and* δ *are given functions of time that are consistent with the initial condition (1.4.10). The other classes of problems stated above can be reformulated in the same way.*

Remark 1.4.2. ***(Problems in unbounded domains)*** *Some problems refer to systems in a half-space* $x \in [0, \infty)$, *or in the whole space* $x \in \mathbb{R}$. *In this case, the boundary conditions have to be stated as above, at the boundaries* $x = 0$ *and* $x \to \infty$, *or for* $x \to -\infty$ *and* $x \to \infty$. *Asymptotic behavior can be studied when the time variable goes to infinity.*

• ***Sufficient initial and boundary conditions.*** The above statement of the problem has been obtained by linking to the evolution equations suitable information on the behavior of the state variable on the boundary of the domain of the dependent variable. The number of boundary conditions must match the highest order of the partial derivative with respect to the dependent variable. If the model is provided with enough initial and boundary conditions to find a solution, the relative mathematical problem is dependent to be ***well formulated.***

Example 1.4.1

Mathematical Problems for the Heat Diffusion Model

Let us consider the heat conduction problem

$$\frac{\partial u}{\partial t} = \kappa \nabla^2 u + f(t, \mathbf{x}; u)\,, \qquad \mathbf{x} \in \mathcal{D}. \tag{1.4.15}$$

This problem is well formulated if it is joined, for instance, to the initial condition

$$u(t=0, \mathbf{x}) = u_0(\mathbf{x})\,, \qquad \mathbf{x} \in \mathcal{D}, \tag{1.4.16}$$

and to the boundary conditions

$$\begin{aligned} u(t, \mathbf{x}) &= u_d(t, \mathbf{x})\ , \qquad \text{if} \ \ \mathbf{x} \in \partial\mathcal{D}_d\,, \\ \mathbf{n}(t, \mathbf{x}) \cdot \nabla u(t, \mathbf{x}) &= u_{\mathbf{n}}(t, \mathbf{x})\ , \qquad \text{if} \ \ \mathbf{x} \in \partial\mathcal{D}_{\mathbf{n}}\,, \end{aligned} \tag{1.4.17}$$

where $\partial\mathcal{D}_d$ and $\partial\mathcal{D}_{\mathbf{n}}$ form a suitable partition of the boundary $\partial\mathcal{D}$ and $\mathbf{n}$ is the normal to $\partial\mathcal{D}_{\mathbf{n}}$.

□

If the model is defined by a system of equations, the initial and boundary conditions have to be assigned for *each* equation according to the rules stated above. For problems in an unbounded domain the same reasoning developed for problems in one space dimension can technically be generalized to problems in more than one space domain.

Remark 1.4.3. *Problems with higher-order derivatives need additional conditions. For instance, for problems involving time derivatives of the order $r \geq 1$, the initial conditions at $t = 0$ have to be prescribed for the dependent variables and their time derivatives up to the order $r - 1$. The same occurs for space derivatives.*

• ***Well-posed problems.*** The notion of well-posed initial-boundary value problems can be stated as follows: *the problem has a solution*, this solution is *unique* and *continuously depends on the various data* of the problem. The data of the problem are referring to the coefficients of the partial differential equation, the functions appearing in boundary and initial conditions, and also the (geometric) region on which the equation has to hold.

In practical situations, the continuity with respect to the problem data means that the solution of the problem changes only slightly when the conditions by which the problem is well formulated change slightly.

The well-posedness is not an immediate consequence of the statement of the problem. It has to be proven by a ***specific qualitative analysis***. Such an analysis requires a detailed knowledge of the theory of partial differential equations, but this is not one of the aims of this book. The reader is referred to the specialized literature; e.g., Lions and Magenes (1969); Evans (1998). Here, it is only pointed out that the study of the well-posedness first requires a precise definition of the class of functions to which the solution belongs. Roughly speaking, the following guidelines apply:

– The auxiliary conditions imposed must not be too many or a solution will not exist.

– The auxiliary conditions imposed must not be too few or the solution will not be unique.

– The kind of auxiliary conditions must be correctly matched to the type of the equation or the solution will not depend continuously on the data.

We provide only a brief analysis of the well-posedness of some models within the framework of the above.

The conditions for well-posed ***hyperbolic equations*** is closely related to the geometry of the ***characteristic curves***. Let us consider for instance the simplest case of linear transport equation on the whole space $\mathbb{R}^3$. The dependent variable $u(t, \mathbf{x})$ is a scalar function which satisfies

$$\frac{\partial u}{\partial t}(t, \mathbf{x}) + \mathbf{b}(\mathbf{x}) \cdot \frac{\partial u}{\partial \mathbf{x}}(t, \mathbf{x}) = 0\,, \quad t > 0, \quad \mathbf{x} \in \mathbb{R}^3 \tag{1.4.18}$$

complemented with the initial condition

$$u(t = 0, \mathbf{x}) = u_0(\mathbf{x})\,, \qquad \mathbf{x} \in \mathbb{R}^3\,,$$

where $\mathbf{b}(\mathbf{x}) = (b_1(\mathbf{x}), b_2(\mathbf{x}), b_3(\mathbf{x}))$ is a given vector-valued function on $\mathbb{R}^3$. Then, the well-posedness of Eq. (1.4.18) is equivalent to the well-posedness of the set of ordinary differential equations:

$$\frac{dX_i(t, \mathbf{x})}{dt} = -b_i(X_1(t, \mathbf{x}); X_2(t, \mathbf{x}); X_3(t, \mathbf{x}))\,,$$

with initial conditions $X_i(0, \mathbf{x}) = x_i$, $i = 1, 2, 3$. One sees that the well-posedness of (1.4.18) is strongly related to the regularity properties of the function $\mathbf{b}(\mathbf{x})$. More difficulties appear if one considers initial-boundary value hyperbolic problems. In particular, one is faced with the problem of compatibility between the behavior of the characteristic curves near the boundaries and the boundary conditions; see Evans (1998), Chapter 3. In general, it is possible to prove a ***local existence result*** for (well-formulated) hyperbolic problems while the global well-posedness occurs generally on open unbounded regions.

There exists a very large variety of analytic tools well suited to prove the well-posedness of both ***parabolic*** and ***elliptic problems***. One may cite, among others, the general theory of pseudo-differential operators based upon the representation of partial differential equations by means of suitable Fourier transforms. Another standard way to construct the solution of a (time-dependent) parabolic problem is to recast the original partial differential equation as a suitable fixed point problem. To prove the existence of a fixed point, compactness methods are a powerful tool.

Finally, we point out that ***variational techniques*** are well suited to prove that both parabolic and elliptic problems are well posed. The variational approach consists in looking for the solution of a given (well-formulated) problem as the minimum of some energy functional over some convex subset of the functional space in which the solution exists. Elliptic problems (with, e.g., mixed boundary conditions) are generally well posed on a sufficiently smooth bounded and open set D of $\mathbb{R}^n$. We stress that, while uniqueness of the solutions to hyperbolic problems is a difficult task, parabolic problems generally lead to a unique solution through maximum principles.

1.5 Guidelines for the Application of Collocation Methods

The interplay between engineering sciences, technology, and applied mathematics often leads to the analysis of initial and/or boundary value problems for nonlinear partial differential equations. Several interesting problems are also stated in terms of integro-differential equations. The preceding section has given an account of the statement of mathematical problems related to models of interest in applied sciences.

A large variety of solution methods can be developed to help solve the above class of problems. The selection of the proper method is, in fact, one of the most difficult tasks of applied mathematics and mainly depends on the structure of the problem to be solved and, to a minor extent, on the aims of the simulation.

A well-known solution technique of nonlinear initial-boundary value problems for nonlinear partial differential equations is the generalized collocation method originally called the ***differential quadrature method***. The original literature can be found in Bellman et al. (1971), (1972).

A concise account concerning the application of the method can be given with reference to the initial-boundary value problem for models described by partial differential equations. The brief description given in what follows is analyzed in detail and generalized in the chapters which follow. Bearing this in mind, the method can be applied as follows.

i) The space variables, say x, y, are discretized into a suitable number of collocation points x_i, y_j.

ii) The dependent variable $u = u(t, x, y)$ is approximated by interpolating polynomials or functions through the values $u_{ij}(t) = u(t, x_i, y_j)$ of the dependent variable in the collocation points. Lagrange polynomials, splines, and sinc functions are usually adopted for the interpolation.

iii) The space derivatives are approximated using the interpolation mentioned in item ii).

iv) The initial-boundary value problem is transformed into an initial value problem for a system of ordinary differential equations that describe the evolution of the values $u_{ij}(t)$ of u in the nodes.

v) Boundary conditions are imposed at the collocation points corresponding to the boundary of the domain of the independent variables.

vi) The solution of the initial-boundary value problem is then obtained by solving the initial value problem mentioned in items iv) and v) and interpolating the solution by using the method given in item ii).

This method discretizes the original continuous model (and problem) into a discrete (in space) and continuous (in time) model, with a finite number of degrees of freedom, while the initial-boundary value problem is transformed into an initial value problem for ordinary differential equations.

A similar method can be developed for the solution of the initial value problem for some classes of integro-differential equations. Again, the initial value problem is transformed into an initial value problem for the values of the dependent variable in the collocation nodes.

This method is well documented in the literature on applied mathematics; it was first proposed by Bellman and Casti (1971) and developed by several authors in deterministic (Satofuka (1983)) and stochastic frameworks, Bellomo and Flandoli (1988). Additional applications and developments are due, among others, to Chen (1999), (2000) in the context of ordinary and partial differential equations, Shu and Richards (1992) concerning the solution of Navier–Stokes equations, and Bert and Malik (1996), (1998) concerning the analysis of the vibration of cylinder shells or rectangular plates.

Let us also mention the papers by Karami and Malekzadeh (2002), (2004) concerning the application of the method to various problems of vibration analysis, and by Artioli, Gould, and Viola (2005) and Artioli and Viola (2005), concerning the solution of mathematical problems in structured mechanics. Finally, the reader can refer to the valuable monograph Shu (1992) that deals with various aspects of the generalized quadrature method in the context of fluid mechanics.

Without discussing, at this stage, the validity of the method, with respect to alternative ones (which can only be done for well-defined problems), it is well documented that the method can provide a useful discretization of continuous models and efficiently deals with nonlinearities including the ones related to implicit boundary conditions. These features make the method interesting for engineering applications, as is documented in the review paper by Bert and Malik (1996) and in the bibliography therein cited. Their paper provides an interesting and detailed report on the application of the original differential quadrature (collocation) method to several engineering problems.

On the other hand, it is known that the method does not generally work in some circumstances. For instance, referring to the Dirichlet problem, the classical Lagrange interpolation is not useful for dealing with problems in unbounded domains or with solutions that are oscillating, with high frequency, in the space variables. This problem can be overcome with a suitable use of sinc functions in the analysis of nonlinear problems, see Lund and Bowers (1992) and Stenger (1983), (1993). A further important aspect is the estimation of the error bounds.

It can be concluded that several studies have been developed to generalize and improve the mathematical method, which has subsequently been applied to interesting engineering problems. These improvements have generated a mathematical method, called the ***generalized collocation method***, which is useful for solving a large class of nonlinear problems in applied sciences.

1.6 Problems

PROBLEM 1.1

Consider the nonlinear transport and diffusion of a chemical (Example 1.2.1) and write it in terms of dimensionless variables.

Hint: Follow the same procedure applied in Examples 1.3.1 and 1.3.2.

PROBLEM 1.2
Identify, in all models proposed in Section 1.2, those which can be classified as parabolic equations.

Hint: Follow the same procedure indicated in the second part of Section 1.4.

PROBLEM 1.3
Identify, in all models proposed in Section 1.2, those which can be classified as hyperbolic equations

Hint: Follow the same procedure indicated in the second part of Section 1.4.

PROBLEM 1.4
Show how the models proposed in Section 1.2 generate, in the steady case, elliptic models.

Hint: Equating the time derivative to zero generates equations which can be discussed following the method of Section 1.4.

PROBLEM 1.5
Consider all models proposed in Section 1.2 by dimensional variables and write them in terms of dimensionless variables.

Hint: Follow the same procedure applied in Problem 1.1.

PROBLEM 1.6
Show that both parabolic and hyperbolic models in the steady case generate elliptic models.

Hint: Follow the same procedure applied in Problem 1.4.

PROBLEM 1.7
Write the statement of mathematical problem (Dirichlet and Neumann) for parabolic models with second-order space derivatives in one space variable.

Hint: Develop this problem in the case of the linear diffusion model.

PROBLEM 1.8
Write the statement of mathematical problem in unbounded domains for wave models.

Hint: Apply the method of Section 1.4.

PROBLEM 1.9

Generalize some of the models of Section 1.2 in two space dimensions considering the case of isotropic and anisotropic material properties.

Hint: The generalization is immediate in the case of isotropic materials. In the case of anisotropic behavior, one has to consider space dependent parameters.

PROBLEM 1.10

Write the mathematical statement of the problems for models in two space dimensions dealt with in Problem 1.9.

Hint: Apply the method described in Section 1.4.

2

Lagrange and Sinc Collocation Interpolation Methods

2.1 Introduction

A brief introduction to the solution of mathematical problems using generalized collocation-interpolation methods has already been given in Chapter 1. As we have seen, the application method requires the development of suitable interpolation techniques of functions of the time and of one or two space variables. These mathematical tools are technically described in this chapter in view of their application to the solution of nonlinear evolution problems described by partial differential and integro-differential equations.

The interpolation method is applied to functions with values known over suitable collocation points of the space variables. It provides a continuous approximation of the functions over the space variables, and allows the calculation of the local derivatives. Considering that the values of the functions in the collocation points may be given as continuous functions of time, the approximation is continuous in time and space.

Several examples and applications provide technical indications to estimate the error bounds both for the function and for its space derivatives. Simulations are obtained by scientific programs developed with the software ***Mathematica***®. The Appendix to this book reports the main features and potential ability of several programs. Specifically, the various programs used in this chapter are reported in Sections A.2–A.6.

The contents are developed in five more sections.

• Section 2.2 shows how *time-dependent functions in one space dimension can be interpolated*, and *approximated*, by using Lagrange polynomials and sinc functions. This section also shows how the interpolation can be applied to approximate first- and higher-order derivatives of given functions. Calculations are developed both in bounded and unbounded domains; it is also shown that the use of Lagrange polynomials appears to be more efficient for interpolation and approximation over bounded domains, while the use of sinc functions appears to be appropriate for functions in unbounded domains.

• Section 2.3 develops a *similar analysis for functions in two space dimensions.* Again Lagrange polynomials and/or sinc functions are used, and it is shown how both types of interpolating functions can be used: one for each space variable. For instance, for problems on a strip, Lagrange polynomials can be used for the bounded space variable, while sinc functions are used for the unbounded variable.

• Section 2.4 presents various examples and applications addressed not only to show the *practical application of the method*, but also to provide useful indications concerning the *selection of the parameters characterizing the method*: number of collocation points and localization of the collocation. This section also provides useful indications on a strategy to select either Lagrange polynomials or sinc functions.

• Section 2.5 develops a *critical analysis on the contents of this chapter* focused on various computational experiments.

• Section 2.6 presents various *problems* to help the reader practice the application of the method.

This chapter is based on the use of Lagrange polynomials or sinc functions to interpolate and approximate functions of time and space. Their application is technically rapid and generally ensures efficient results. On the other hand, the reader should be aware that other types of interpolations can be used, such as splines. This technical development is left to the reader's initiative; however, a brief discussion on the use of orthogonal polynomials is given in the last chapter of the book.

2.2 Collocation Methods in One Space Dimension

This section deals with interpolation techniques of functions of time and one dependent space variable using Lagrange polynomials and sinc functions. The method is presented in view of the solution to nonlinear boundary value problems such that the dependent variables depend on time and one space variable.

Let us consider functions of the type

$$u = u(t,x), \quad [0,1] \times [0,1] \to [0,1], \tag{2.2.1}$$

where both independent and dependent variables, as shown in Chapter 1, take values in the interval $[0,1]$ after having been normalized with respect to suitable reference quantities. The domain of the space variable and/or of the dependent variable can also be $[-1,1]$.

The above normalization is possible when the real space variable x_r is defined over a bounded interval, say $x_r \in [x_m, x_M]$. Otherwise, if x_r is defined over the whole real line (or over the positive half-line), then one may either let $x \in (-\infty, \infty)$ (or $x \in [0,\infty)$), or apply a change of variable, say $z = z(x)$ with z chosen in such a way that $z \in [0,1]$.

Obviously the time variable spans over an arbitrary interval subset of $\mathbb{R}_+$, say over $[0,T]$ $(T > 0)$. However, once the time T necessary for the simulations has been fixed, the scaled time is then obtained by dividing the real time by T, so that $t \in [0,1]$. Additional reasoning will be referred to specific examples.

Consider then functions $u = u(t,x)$ defined over $[0,1] \times [0,1]$, such that u defines a one-to-one map from $x \in [0,1]$ into the domain of u, for every fixed $t \in [0,1]$, and let us define the collocation

$$I_x = \{x_1 = 0, \ldots, x_i, \ldots, x_n = 1\}. \tag{2.2.2}$$

This collocation can be either equally spaced

$$x_i = (i-1)h, \qquad h = \frac{1}{n-1}, \qquad i = 1, \ldots, n, \tag{2.2.3}$$

or identified by a ***Chebychev-type collocation*** with decreasing values of the measure $|x_i - x_j|$ towards the borders:

$$x_i = \frac{1}{2} - \frac{1}{2}\cos\left(\frac{i-1}{n-1}\pi\right), \qquad i = 1, \ldots, n. \tag{2.2.4}$$

In general, $u = u(t,x)$ can be interpolated and approximated by means of Lagrange polynomials as follows:

$$u(t,x) \cong u^n(t,x) = \sum_{i=1}^{n} L_i(x) u_i(t), \tag{2.2.5}$$

where $u_i(t) = u(t, x_i)$, and the Lagrange polynomials are given by the expression

$$L_i(x) = \frac{(x - x_1)\dots(x - x_{i-1})(x - x_{i+1})\dots(x - x_n)}{(x_i - x_1)\dots(x_i - x_{i-1})(x_i - x_{i+1})\dots(x_i - x_n)} \cdot \tag{2.2.6}$$

Since $L_i(x_j) = \delta_{ij}$ one has $u^n(t, x_i) = u_i(t)$. Therefore the function $u = u(t, x)$, of time and space, is approximated by a collection of n profiles $\{u_i(t)\}$ for $i = 1, \dots, n$.

This interpolation can be used to approximate the partial derivatives of the function u in the nodal points of the discretization

$$\frac{\partial^r u}{\partial x^r}(t; x_i) \cong \sum_{h=1}^{n} a_{hi}^{(r)}(n) u_h(t), \qquad r = 1, 2, \dots, \tag{2.2.7}$$

where

$$a_{hi}^{(r)}(n) = \frac{d^r L_h}{dx^r}(x_i) \cdot \tag{2.2.8}$$

The values of the coefficients depend on the number of collocation points n and on the type of collocation. Technical calculations provide the following result:

$$h \neq i: \quad a_{hi}^{(1)} = \frac{\prod(x_i)}{(x_i - x_h)\prod(x_h)}, \qquad a_{ii}^{(1)} = \sum_{h \neq i} \frac{1}{x_i - x_h}, \tag{2.2.9a}$$

and

$$h = i: \quad a_{ii}^{(1)} = \sum_{h \neq i} \frac{1}{x_i - x_h}, \tag{2.2.9b}$$

where

$$\prod(x_i) = \prod_{p \neq i}(x_i - x_p), \qquad \prod(x_h) = \prod_{p \neq h}(x_h - x_p). \tag{2.2.10}$$

Higher-order coefficients may be computed exploiting the following recurrence formula:

$$a_{hi}^{(r)} = r\left(a_{hi}^{(1)}\, a_{ii}^{(r-1)} + \frac{a_{hi}^{(r-1)}}{x_h - x_i}\right), \qquad a_{ii}^{(r+1)} = -\sum_{h \neq i} a_{hi}^{(r+1)}. \tag{2.2.11}$$

Remark 2.2.1. *It is well known that accuracy of the interpolation can be obtained by the selection of the proper collocation points to be related to the selection of the interpolation functions; see Bellomo and Preziosi (1996). In the case of Lagrange interpolation, a Chebychev collocation is needed. Indeed, the examples in Section 2.4 show how using collocation with equal spacing generally gives very disappointing results.*

In the more general case, when x belongs to an unbounded interval, it is still possible to use (2.2.5) and (2.2.6). In fact, let us suppose that $x \in (-\infty, \infty)$ for functions which tend to zero at infinity:

$$\lim_{|x| \to \infty} u(t, x) = 0\,, \quad \forall\, t \geq 0\,. \tag{2.2.12}$$

Then, the following change of variable can be applied:

$$z = \frac{e^x}{1 + e^x}\,, \tag{2.2.13}$$

so that $z = 1$ when $x \to \infty$, and $z = 0$ when $x \to -\infty$, and the inverse mapping is as follows:

$$x = \log \frac{z}{1 - z} \,\cdot \tag{2.2.14}$$

If $x \in [0, \infty)$ for functions which tend to zero at infinity, the following change of variable can be applied:

$$z = 1 - e^{-x}\,, \tag{2.2.15}$$

which is such that $z = 0$ when $x \to 0$, while $z = 1$ when $x \to \infty$, so that the inverse mapping is given by

$$x = -\log(1 - z)\,. \tag{2.2.16}$$

Similar calculations can be developed in the case of sinc functions which are naturally defined for the interval $x \in (-\infty, \infty)$:

$$u(t, x) \cong u^n(t, x) = \sum_{i=-n}^{n} S_i(x; h) u_i(t)\,, \tag{2.2.17}$$

where

$$S_i(x; h) = \frac{\sin z_i}{z_i}\,, \qquad z_i = \frac{\pi}{h}(x - ih)\,. \tag{2.2.18}$$

In this case the collocation is as follows:

$$x_n = hn \quad \text{for any relative integer } n \in \mathbb{Z}\,, \tag{2.2.19}$$

where h may be related to a characteristic length of the physical system. Hence, sinc functions require an equally spaced collocation.

The above interpolation also operates for functions defined over the interval $[0,1]$, by assuming that

$$u(t,x) = 0 \quad \text{whenever} \quad x \notin [0,1], \quad \forall t \geq 0\,. \tag{2.2.20}$$

Technical calculations, see Bonzani (1997), yield, for any integer r, a simple expression of the coefficients of the space derivative matrices:

$$\begin{cases} a_{ji}^{(2r)} = \dfrac{(-1)^{i-j}}{h^{2r}(i-j)^{2r}} \displaystyle\sum_{k=0}^{r-1} (-1)^{k+1} \frac{2r!}{(2k+1)!} \pi^{2k} (i-j)^{2k}, \\[2ex] a_{ji}^{(2r+1)} = \dfrac{(-1)^{i-j}}{h^{2r+1}(i-j)^{2r+1}} \displaystyle\sum_{k=0}^{r} (-1)^{k+1} \frac{(2r+1)!}{(2k+1)!} \pi^{2k} (i-j)^{2k}\,, \end{cases} \tag{2.2.21a}$$

for $i \neq j$ and

$$a_{ii}^{(2r)} = \left(\frac{\pi}{h}\right)^{2r} \frac{(-1)^r}{(2r+1)}\,, \qquad a_{ii}^{(2r+1)} = 0\,, \tag{2.2.21b}$$

for $i = j$.

The computation of the space derivatives in the collocation points generates an $n \times n$ matrix of coefficients, where n is the number of collocation nodes. The continuous approximation of the space derivatives can be obtained using the same interpolation formula used for the functions. We stress that the interpolation is exactly satisfied in the nodal points, while partial space derivatives are only approximated.

Finally, let us remark that the choice of the interpolating functions is not limited to Lagrange polynomials and sinc functions. In general, it is possible to look for interpolations of the type

$$u(t,x) \cong u^n(t,x) = \sum_{i=1}^{n} \chi_i(x) u_i(t)\,, \tag{2.2.22}$$

where $\chi_i(x_i) = 1$, and $\chi_i(x_j) = 0$ whenever $i \neq j$. Here, χ represents a general fundamental interpolating function, such as Lagrange polynomials or sinc functions, but also fundamental splines, Legendre polynomials with Legendre collocation points, and so on. The presentation that follows refers, for simplicity, to the interpolations (2.2.6) and (2.2.18) leaving further generalizations to the reader, who may take advantage of the general contents of the book by Bellomo and Preziosi (1996).

2.3 Collocation and Interpolation in Two Space Dimensions

The method described in Section 2.2 can be technically developed for time-dependent functions in two space variables: $u = u(t, x, y)$. Let us consider first functions defined over the domain $[0,1] \times [0,1] \times [0,1]$, such that u defines a one-to-one map from $D = [0,1] \times [0,1]$ into the domain of u, for every fixed $t \in [0,1]$. Moreover, let us introduce, in addition to the collocation I_x, the following collocation for the y-variable:

$$I_y = \{y_1 = 0, \dots, y_j, \dots, y_m = 1\}, \quad j = 1, \dots, m. \tag{2.3.1}$$

It follows that

$$u = u(t, x, y) \cong u^{nm}(t, x, y) = \sum_{i=1}^{n} \sum_{j=1}^{m} L_i(x) L_j(y) u_{ij}(t), \tag{2.3.2}$$

or

$$u = u(t, x, y) \cong u^{nm}(t, x, y) = \sum_{i=1}^{n} \sum_{j=1}^{m} S_i(x; h) S_j(y; h) u_{ij}(t). \tag{2.3.3}$$

In principle, mixed-type interpolations can also be used:

$$u = u(t, x, y) \cong u^{nm}(t, x, y) = \sum_{i=1}^{n} \sum_{j=1}^{m} L_i(x) S_j(y; h) u_{ij}(t), \tag{2.3.4}$$

or

$$u = u(t, x, y) \cong u^{nm}(t, x, y) = \sum_{i=1}^{n} \sum_{j=1}^{m} S_i(x; h) L_j(y) u_{ij}(t). \tag{2.3.5}$$

The use of a mixed-type interpolation is useful if one of the two variables is defined over the whole real line. For instance, let us suppose that $x \in [0,1]$ and $y \in (-\infty, \infty)$; then the collocation

$$y_m = hm \qquad \text{for any relative integer } m \in \mathbf{Z}\,, \tag{2.3.6}$$

can be used for the y variable, so that the interpolation is as follows:

$$u = u(t,x,y) \cong u^{nm}(t,x,y) = \sum_{i=1}^{n} \sum_{j=-m}^{m} L_i(x) S_j(y;h) u_{ij}(t)\,. \tag{2.3.7}$$

The approximation of the space derivatives in the collocation points is obtained by calculations analogous to those of the one-dimensional case. When the space variables are defined over the domain $D = [0,1] \times [0,1]$, one has

$$\frac{\partial^r u}{\partial x^r}(t; x_i, y_j) \cong \sum_{h=1}^{n} a_{hi}^{(r)} u_{hj}(t)\,, \tag{2.3.8}$$

and

$$\frac{\partial^r u}{\partial y^r}(t; x_i, y_j) \cong \sum_{k=1}^{m} b_{kj}^{(r)} u_{ik}(t)\,, \tag{2.3.9}$$

while mixed-type derivatives are given by

$$\frac{\partial^2 u}{\partial x \partial y}(t; x_i, y_j) \cong \sum_{h=1}^{n} \sum_{k=1}^{m} a_{hi}^{(1)} b_{kj}^{(1)} u_{hk}(t)\,, \tag{2.3.10}$$

where the coefficients a and b are given by the expressions reported in Section 2.2.

Also, for two space dimensions the interpolation is exactly satisfied in the nodal points

$$u_{ij}(t) = u^{nm}(t, x_i, y_j)\,, \tag{2.3.11}$$

while partial derivatives are only approximated.

If the mathematical problem in two space dimensions is not defined on rectangular domains, the collocation has to be dealt with referring to the geometry of the domain of the space variables, while the interpolation method needs to be technically modified. For instance, when the domain is convex with respect to both axes and regular, then the number of collocation points along one of the axes may depend on the collocation on the other axis. This matter needs to be technically treated with reference to specific applications such as those in Chapter 5.

The same technique can also be developed for time-dependent functions in three space variables. However, this specific problem is not discussed here. Its application to the solution of partial differential equations is not practical since it generates an excessively large number of equations.

2.4 Examples and Applications

Some simple applications will be proposed in this section to show some practical aspects of the interpolation method described in Sections 2.2 and 2.3. The examples are all organized to interpolate, by means of a finite number of collocation points, functions which are known analytically. This allows an immediate technical estimate of the approximation error, that is the L_∞-distance between the real function, say $u = u(x)$ and the interpolating-approximating one $u^n = u^n(x)$.

The error, which depends on the number of collocation points, is defined as follows:

$$\mathcal{E}^n = ||u - u^n||_\infty = \sup_{x \in D} |u - u^n| \,. \tag{2.4.1}$$

Analogous estimates can be obtained for the space derivatives:

$$\mathcal{E}^n_x = ||u_x - u^n_x||_\infty = \sup_{x \in D} |u_x - u^n_x| \,, \tag{2.4.2}$$

and

$$\mathcal{E}^n_{xx} = ||u_{xx} - u^n_{xx}||_\infty = \sup_{x \in D} |u_{xx} - u^n_{xx}| \,, \tag{2.4.3}$$

where the subscripts refer to the first- and second-order space derivatives. Analogous definitions can be given for higher-order derivatives.

In general, one should expect $\mathcal{E}^n \to 0$ as $n \to \infty$. On the other hand, a monotone decreasing of the error, say

$$\left(\mathcal{E}^n\right)_n \text{ is a decreasing sequence,} \tag{2.4.4}$$

although often observed in numerical simulations, can be proved only for some special cases.

The same reasoning can be generalized to $\mathcal{E}^n_x$ and $\mathcal{E}^n_{xx}$ and possibly to higher-order derivatives.

The understanding of the mathematical method can be enlarged to analyze the average error identified by the L_1-distance:

$$\mathcal{E}_1^n = \frac{1}{\text{meas}(D)} \int_D |u - u^n|(t,x)dx \,.$$

Analogous expressions can be written for the space derivatives.

For the applications that follow, $D = [0,1]$ and $\text{meas}(D) = 1$. The first example aims to analyze the need of a Chebychev collocation.

Example 2.4.1
Lagrange Interpolations

This example refers to the study of the difference between equally spaced and Chebychev collocations for Lagrange interpolation, and is illustrated for the following function:

$$u = u(x) = \tanh[5(2x-1)] + \frac{1}{5}\sin[5\pi(2x-1)]\,, \quad x \in [0,1]\,. \tag{2.4.5}$$

□

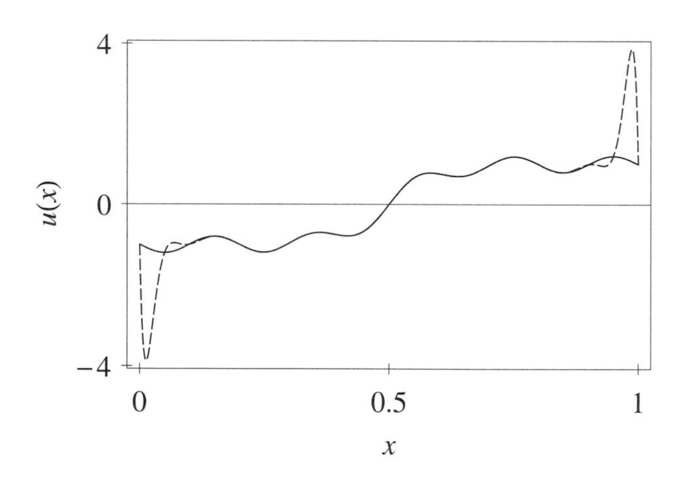

Figure 2.4.1 - Lagrange interpolation of (2.4.5) with 21 collocation points: Chebychev collocation (continuous line), equally spaced collocation (dashed line).

Simulations are shown in Figure 2.4.1 referring to the approximation of the wavy perturbation of the hyperbolic tangent (2.4.5) by Lagrange polynomials with 21 nodes. The continuous line refers both to the analytical function and its interpolation by Chebychev collocation, while the dashed line corresponds to equally spaced nodes. This example clearly shows that the Lagrange interpolation needs a Chebychev collocation: indeed, the error generated by the equally spaced collocation is remarkable, whereas the figure does not show any discrepancy for the analytic curve and its Lagrange interpolation with a Chebychev collocation.

The computations have been performed using the program ***OneDLag*** reported in the Appendix, Section A.2.

The second application provides a comparison between the Lagrange and the sinc interpolations.

Example 2.4.2
Lagrange and Sinc Interpolation

This application compares the Lagrange and sinc interpolations—using the Chebychev collocation and equally spaced collocation, respectively.

The example refers to the interpolation of the function

$$v = v(x) = \exp[-50(2x-1)^2]\,, \quad x \in [0,1]\,. \tag{2.4.6}$$

The first and the second derivatives of the interpolations are also computed, while the errors $\mathcal{E}_x^n$ and $\mathcal{E}_{xx}^n$ are evaluated with respect to the number of nodes. Computations have been performed with the help of the program ***OneDLaSiInt***, see Appendix, Section A.3.

□

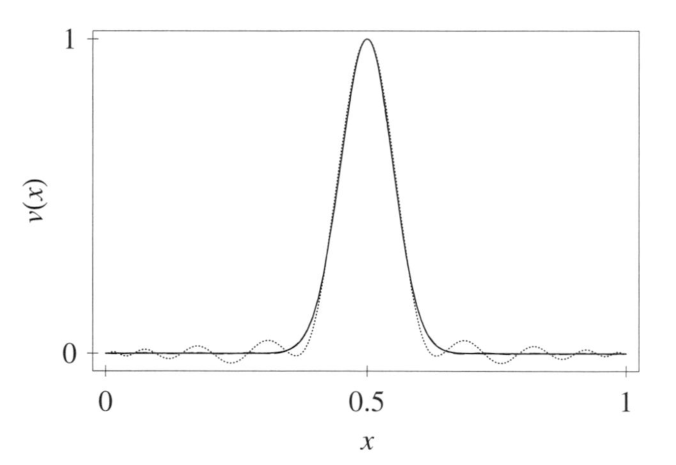

Figure 2.4.2.a - Interpolation, with 21 nodes, of a Gaussian-like function (continuous line): Lagrange polynomials (dashed line) and sinc functions (dotted line). The dashed line practically overlaps with the continuous line of the exact function.

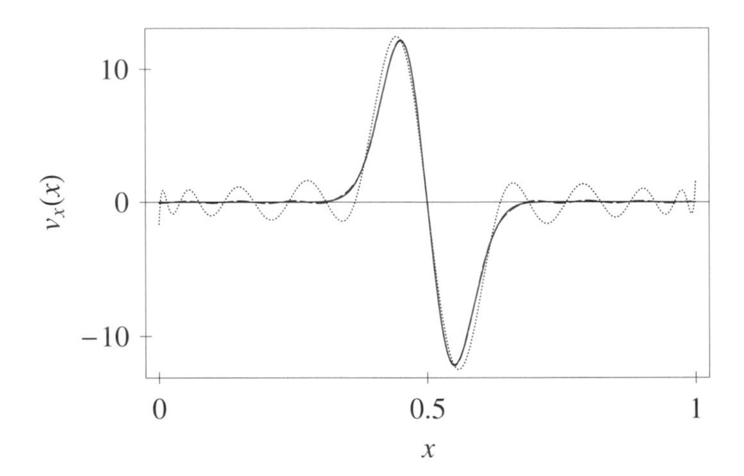

Figure 2.4.2.b - Interpolation of the first derivative of a Gaussian-like function (continuous line) by 21 collocation points.

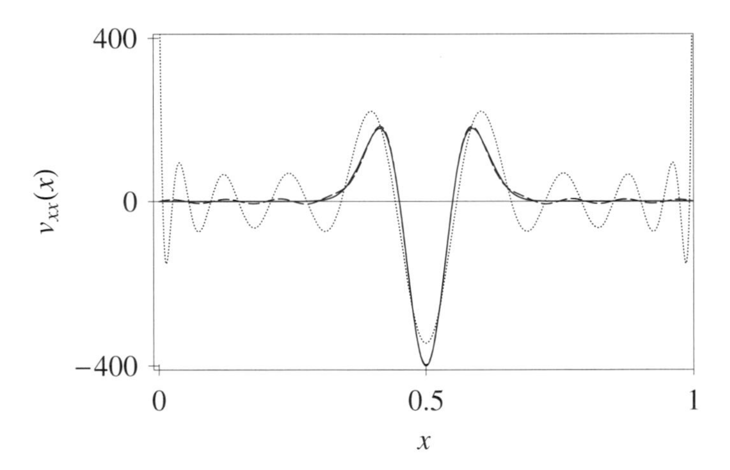

Figure 2.4.2.c - Interpolation of the second derivative of a Gaussian-like function (continuous line) by 21 collocation points.

Figures 2.4.2.a,b,c, show that a relatively more accurate approximation is obtained by Lagrange interpolation (dashed line). This feature is evident in Figure 2.4.2.a when the analytical function (solid line) becomes flat; the interpolation by sinc functions (dotted line) is wavy while the Lagrange interpolation curve is flat. Figures 2.4.2.b and 2.4.2.c further emphasize this aspect related to the errors of the first and second derivatives.

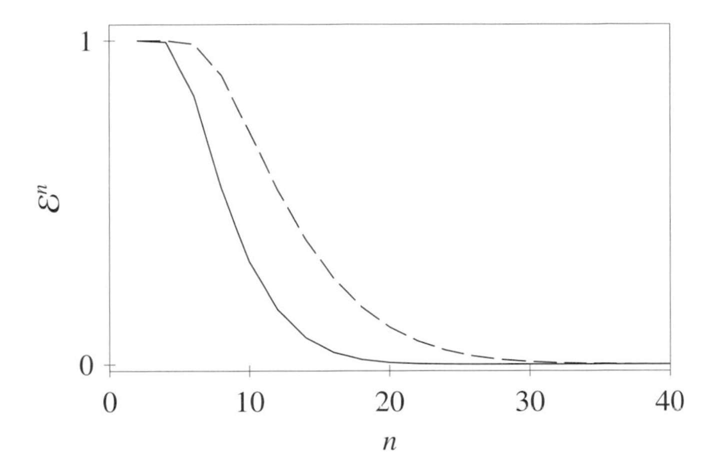

Figure 2.4.2.d - Error by L_∞-norm versus even number of nodes n: Sinc functions (continuous line) and Lagrange polynomials (dashed line).

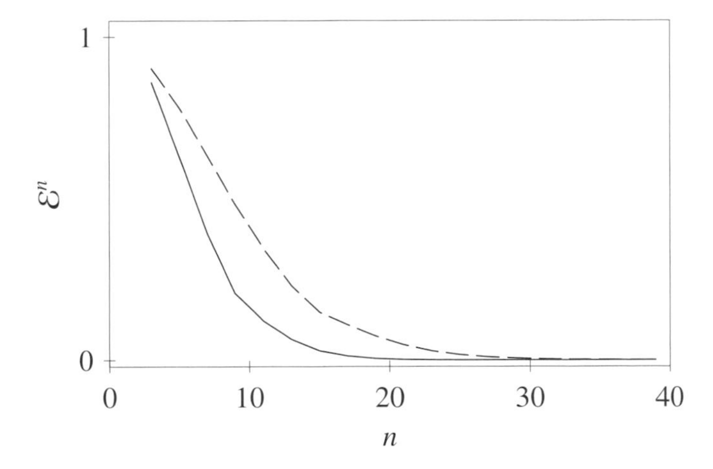

Figure 2.4.2.e - Error by L_∞-norm versus odd number of nodes n: Sinc functions (continuous line) and Lagrange polynomials (dashed line).

Figures 2.4.2.d and 2.4.2.e report the error $\mathcal{E}^n$ corresponding to the sinc interpolation (solid line) and Lagrange interpolation (dashed line), for an even and odd number of nodes, respectively. The errors visualized in Figures 2.4.2.d,e confirm that the best results are obtained in this example by Lagrange interpolation and show the fast decrement of the errors when the nodes number is increasing: when n is in the range 20–30, the errors practically tend to zero.

Example 2.4.3
Error Computation for Example 2.4.1

Consider the interpolation of a function defined as an oscillating perturbation superposed to the hyperbolic tangent (2.4.5). The function is interpolated by Lagrange polynomials and sinc functions (always with Chebychev collocation and equispaced collocations, respectively) making use of the program ***OneDLSerr*** (Appendix: Section A.4).

□

Figure 2.4.3.a visualizes the approximation of the function u given by (2.4.5) (solid line) by sinc functions (dotted line) and Lagrange polynomials with Chebychev collocation (dashed line), while Figures 2.4.3.b and 2.4.3.c visualize the behavior of its first derivative u_x and its second derivative u_{xx} respectively. Figures 2.4.3.d and 2.4.3.e visualize the L_∞-norm error $\mathcal{E}^n$ with an even (solid line) and odd (dashed line) number of nodes for sinc and Lagrange interpolation, respectively, while Figures 2.4.3.f and 2.4.3.g show the L_1-norm error $\mathcal{E}_1^n$ with an even (solid line) and odd (dashed line) number of nodes for sinc and Lagrange interpolation, respectively.

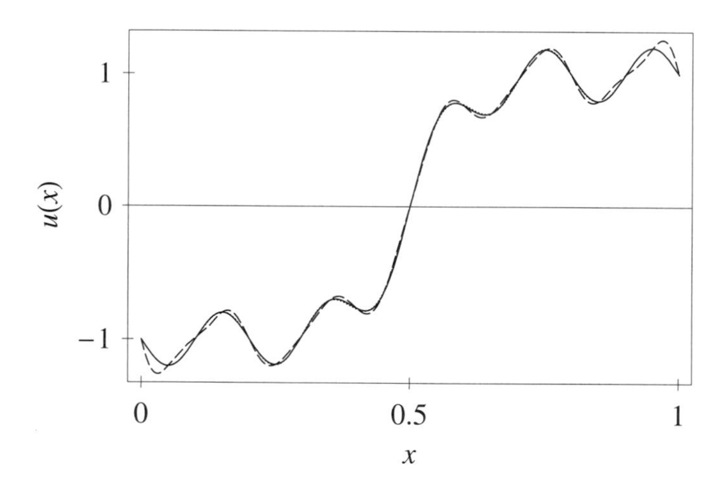

Figure 2.4.3.a - Interpolation by Lagrange polynomials (dashed line) and sinc functions (dotted line), with 21 collocation points, of (2.4.5) (continuous line).

Unlike the previous example, now the behavior of the sinc interpolation is relatively more accurate than the one obtained by Lagrange interpolation. The reason is the behavior of the sinc function which allows us to capture the oscillations of the function $u(x)$, in particular near the boundaries, as testified by the behavior of its derivatives. As also shown in Example 2.4.2,

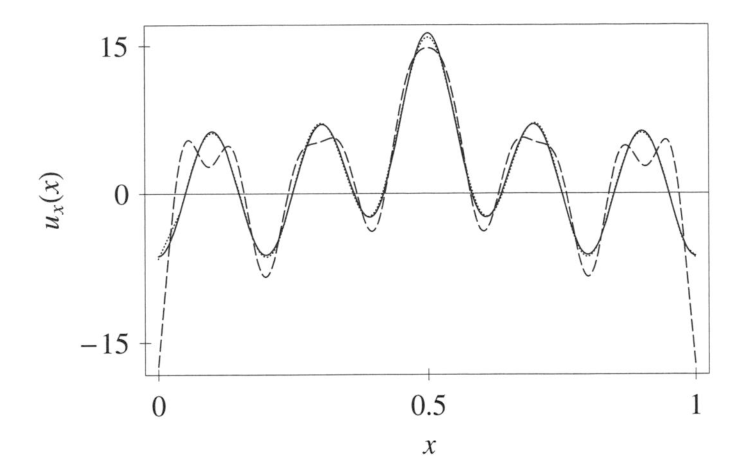

Figure 2.4.3.b - Interpolation by Lagrange polynomials (dashed line) and sinc functions (dotted line), with 21 collocation points, of the first derivative of (2.4.5) (continuous line).

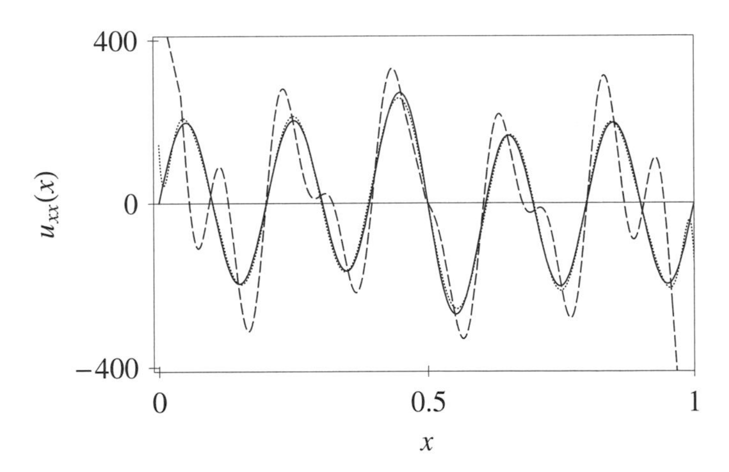

Figure 2.4.3.c - Interpolation by Lagrange polynomials (dashed line) and sinc functions (dotted line), with 21 collocation points, of the second derivative (2.4.5) (continuous line).

in this case the error decreases very fast with an increasing number of nodes, even if the Lagrange interpolation shows a residual error which is significantly different from zero.

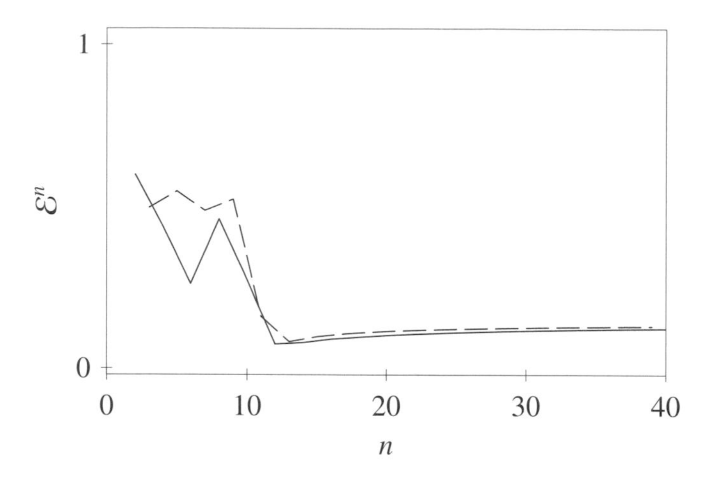

Figure 2.4.3.d - Error ε^n versus n of the interpolation by sinc functions of (2.4.5): even number of nodes (continuous line) and odd number of nodes (dashed line).

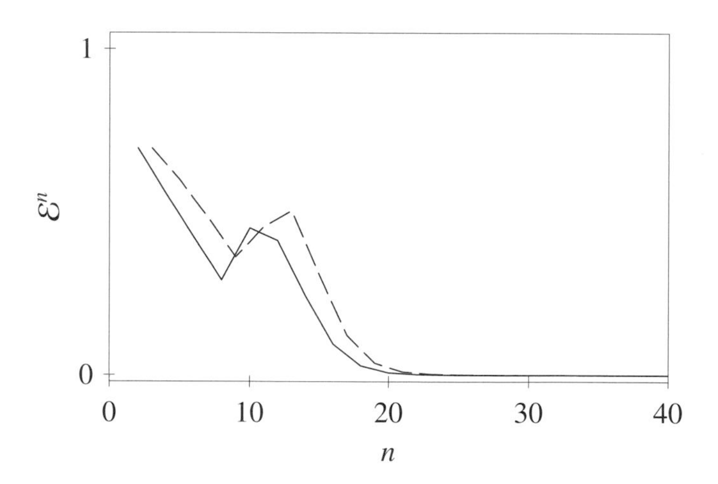

Figure 2.4.3.e - Error ε^n versus n of the interpolation by Lagrange polynomials of (2.4.5): even number of nodes (continuous line) and odd number of nodes (dashed line).

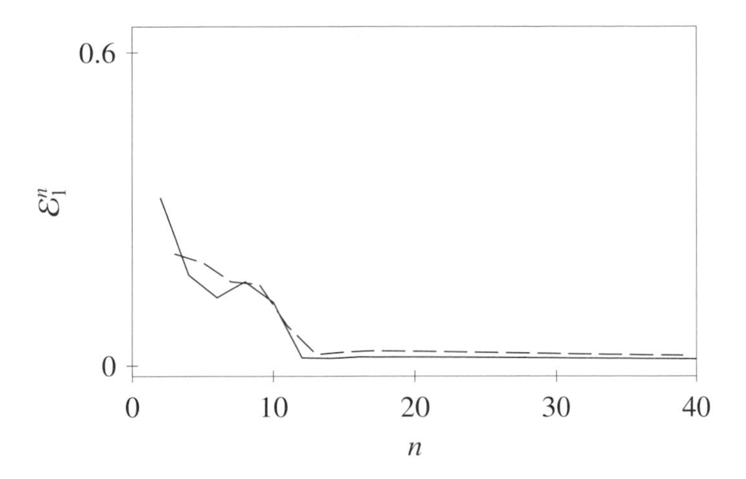

Figure 2.4.3.f - Error ε_1^n versus n of the interpolation by sinc functions of (2.4.5): odd number of nodes (dashed line) and even (continuous line).

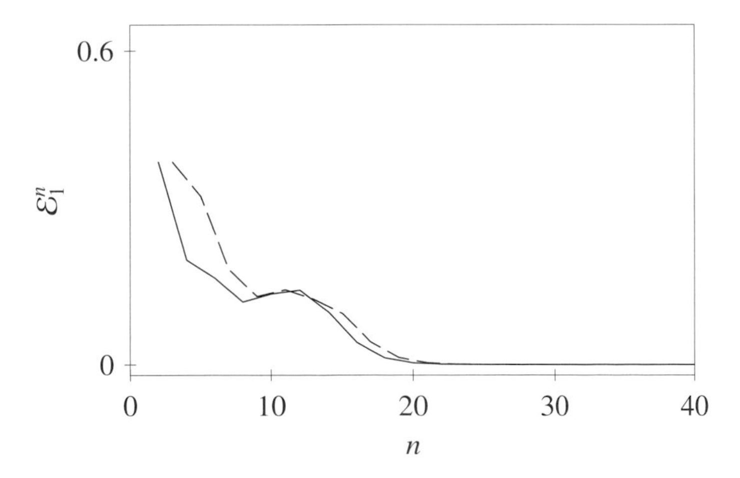

Figure 2.4.3.g - Error ε_1^n versus n of the interpolation by Lagrange polynomials of (2.4.5): odd number of nodes (dashed line) and even (continuous line).

Example 2.4.4
Gaussian Function in Two Space Dimensions

This example refers to the interpolation in two space dimensions, where the test function is the following Gaussian-type function:

$$u = u(x, y) = \exp[-25(2x - 1)^2 - 25(2y - 1)^2], \qquad x, y \in [0, 1]. \quad (2.4.7)$$

Computations have been obtained by programs ***TwoDLaSiInt*** and ***TwoDLaSiErr*** (see Appendix, Sections A.5 and A.6).

□

Figures 2.4.4.a and 2.4.4.b visualize, respectively, the approximation of the function (2.4.7) by sinc functions and the associated error $\mathcal{E}^n \times 10^5$, while Figures 2.4.4.c and 2.4.4.d show, respectively, the approximation of (2.4.7) by Lagrangian polynomials with Chebychev collocation and the corresponding error $\mathcal{E}^n \times 10^3$.

The L_∞-norm error $\mathcal{E}^n$ and the L_1-norm error $\mathcal{E}_1^n$, respectively, with an odd number of nodes for sinc (solid line) and Lagrange interpolation (dashed line) are visualized in Figures 2.4.4.e and 2.4.4.f.

A visual inspection of Figures 2.4.4.a,c and the magnitude of the error show clearly that both Lagrange and sinc interpolations provide very accurate results. However, due to the flat shape of the function $u(x, y)$ in a large part of the domain, and coherently with Example 2.4.2, the Lagrange interpolation leads to a relatively more accurate approximation.

2.5 Critical Analysis

This chapter has shown how the interpolation and approximation of time-dependent functions in one or two space variables can be performed by using Lagrange polynomials and sinc functions. Some applications have been proposed choosing various analytical functions and interpolating them, so that an immediate estimate of the related approximation errors is possible.

The various applications have shown that the approximation by Lagrange polynomials provides accurate results only in the case of Chebychev collocations. On the other hand, there are no well-defined rules to choose between one type of interpolation, between Lagrange or sinc. A heuristic indication is that the use of sinc functions may give relatively more accurate results than Lagrange interpolations for a large number of nodes. On the other hand, if the number of nodes is small, then Lagrange polynomials are generally more efficient. This is an important issue considering that the

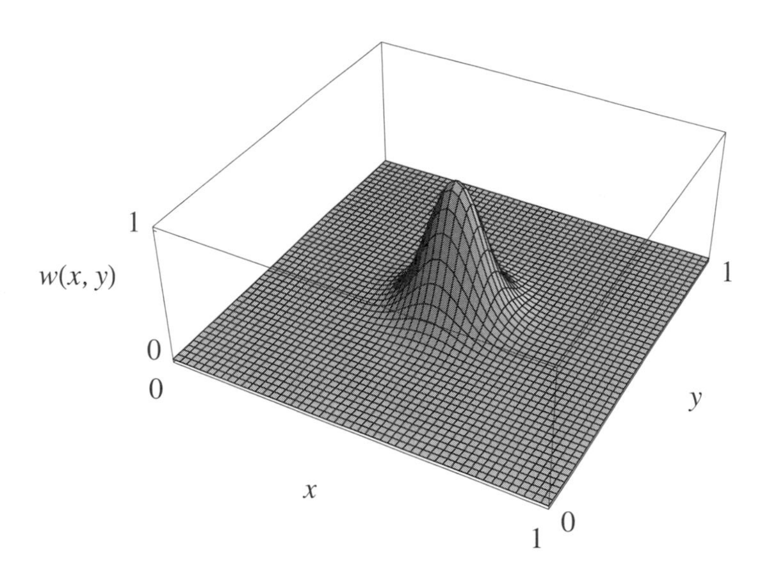

Figure 2.4.4.a - Approximation by sinc functions with 15×15 nodes of (2.4.7).

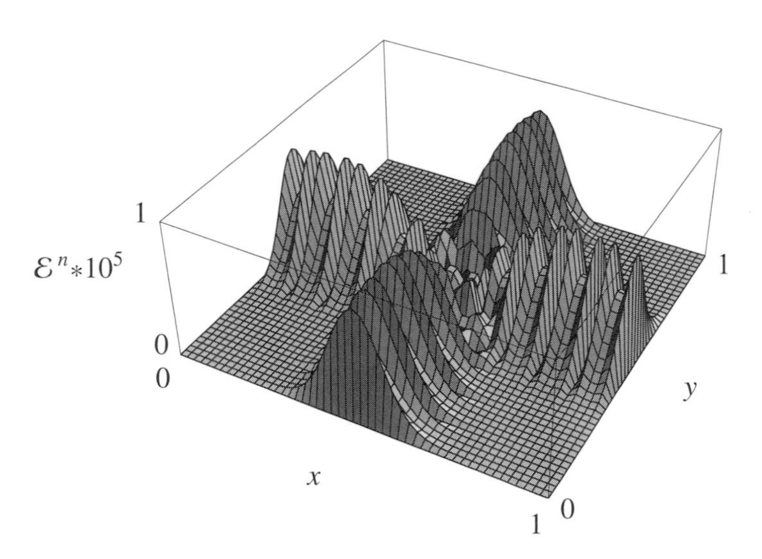

Figure 2.4.4.b - ε in the approximation by sinc functions with 15×15 nodes of (2.4.7).

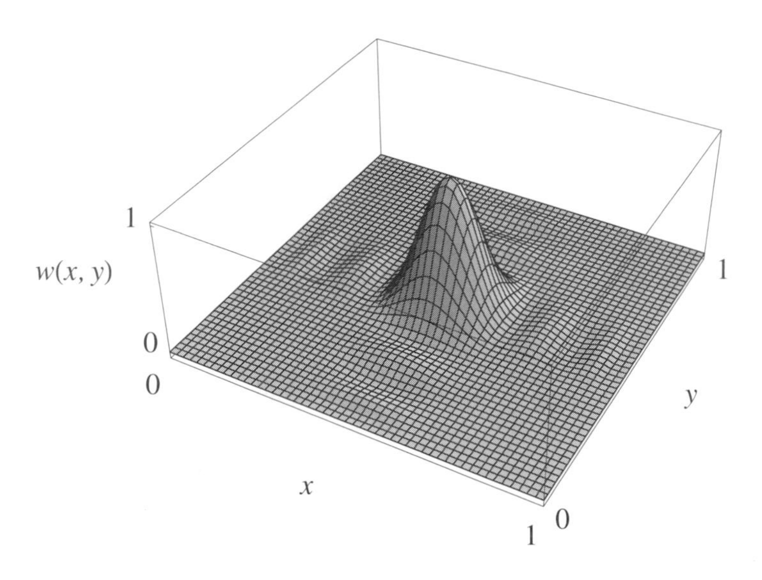

Figure 2.4.4.c - Approximation by Lagrange polynomials with 15×15 nodes of (2.4.7).

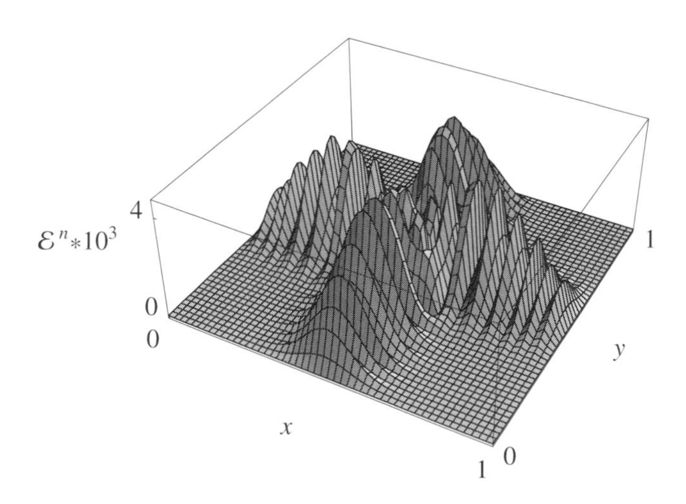

Figure 2.4.4.d - ε in the interpolation by Lagrange polynomials, with 15×15 nodes, of the Gaussian-like function (2.4.7).

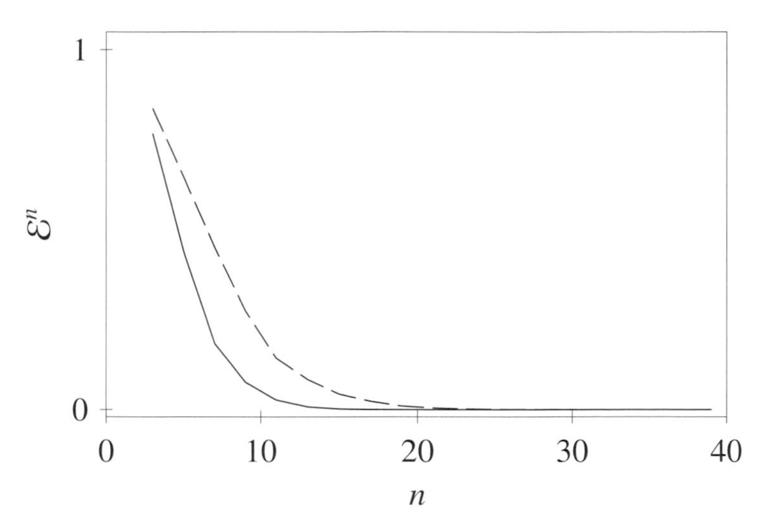

Figure 2.4.4.e - ε^n versus odd number of nodes n in the interpolation of (2.4.7): Sinc functions (continuous line) and Lagrange polynomials (dashed line).

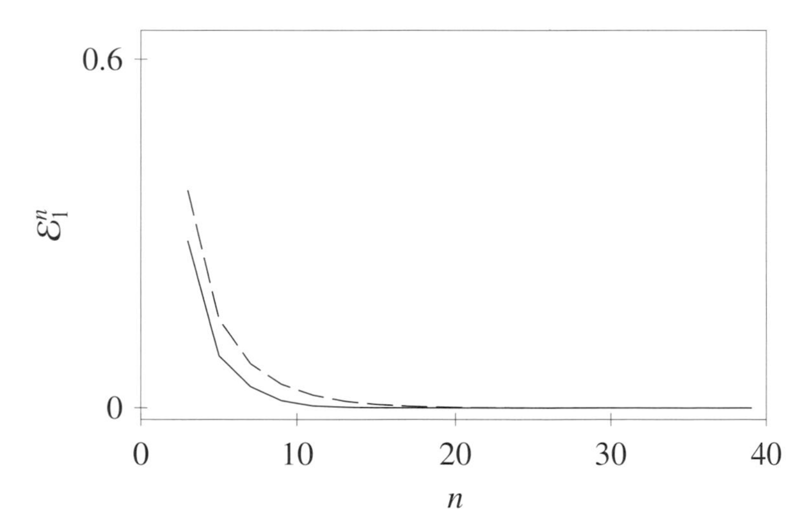

Figure 2.4.4.f - L_1-norm versus odd number of nodes n: Sinc functions (continuous line) and Lagrange polynomials (dashed line).

application of generalized collocation methods requires the solution of a system of ordinary differential equations equal to that of the number of nodes. Therefore, the lower the number of nodes, the less stiff the system of differential equations. Moreover, for functions that are flat at the boundary of the domain, Lagrange interpolation provides more accurate results than the sinc method, as illustrated in Example 2.4.4.

The applications have also shown that the error tends to zero as the number of interpolation nodes tends to infinity, and that the error is monotonically decreasing with different behaviors for an even or odd number of nodes. This consistency result is certainly useful towards the solution of initial-boundary value problems, although it is clear that the number of collocation points cannot be made arbitrarily large—as already mentioned, it would generate an excessively large number of ordinary differential equations.

Various books and papers report a priori estimates of the error bounds as documented in the review by Stenger (1983) or in the paper by Bellomo et al. (2001). This matter is not dealt with in this chapter because the experimental (computational) approach initiated here is relatively more practical for applications. Indeed, the main point is the selection of the best number of nodes related to the fact that, by increasing the number of collocation points, one reduces the interpolation error but increases (as we shall see) the error related to the time integration. An optimal strategy can be developed by computational experiments which will be practically proposed in the next chapters.

Finally, the choice of Lagrange polynomials and sinc functions is related to the fact that the method is based on ***approximation*** generated by ***interpolation***. The alternative of using Bernstein polynomials is also possible; however, the authors' experience suggests limiting the use to the selection proposed in this chapter. On the other hand, if the approximation is obtained by orthogonal functions, the method needs to be technically developed as reported in Chapter 6. In this case, the use of wavelets can be considered. Pertinent information can be found in the books by Daubechies (1992), Meyer and Ryan (1993), and Cattani and Rushchitsky (2007). The general framework is that of spectral approximation, Gottelieb and Orszag (1977).

2.6 Problems

PROBLEM 2.1

Consider the function defined in Example 2.4.2 and compute, for $x \in [0,1]$, the distance between u_x and u_x^n obtained by Lagrange interpolation with Chebychev collocation using 11 nodes. Discuss this result with respect to the one of Example 2.4.3.

Hint: Use the same programs of Examples 2.4.1 and 2.4.3.

PROBLEM 2.2

Consider the same interpolation of Problem 2.1 computing now the distance between u_{xxx} and u^n_{xxx}. Develop a similar analysis for the fourth-order derivatives and discuss the result.

Hint: The distance between the analytic function and the interpolating one increases with increasing order of the derivative.

PROBLEM 2.3

Consider the Gaussian-type function defined by Eq. (2.4.7) and compute the errors $\mathcal{E}^n_x$ and $\mathcal{E}^n_{xx}$ obtained by a sinc interpolation using 20 nodes.

Hint: Use the same program applied in Example 2.4.2.

PROBLEM 2.4

Develop the same analysis of Problem 2.3 using a Lagrange collocation with 20 nodes and a Chebychev collocation.

Hint: Use the same program applied in Example 2.4.2.

PROBLEM 2.5

Develop the same calculations of Problems 2.3 and 2.4 with a variable number of nodes and discuss the results of the computation.

Hint: Use the same procedure of Example 2.4.2.

PROBLEM 2.6

Consider Example 2.4.4 and develop the same analysis of the above problems by using mixed-type interpolations: sinc in the x-axis and Lagrange on the y-axis (and vice versa). Then, discuss the result of the computation.

Hint: Use the same program applied in Example 2.4.2.

PROBLEM 2.7

Consider Problem 2.6 and develop the same analysis concerning the estimates of the approximation for the first partial derivatives.

Hint: Use the same program applied in Example 2.4.2.

PROBLEM 2.8

Develop the same analysis of Problem 2.7 for second-order derivatives.

Hint: Use the same program applied in Example 2.4.2.

PROBLEM 2.9

Select a function $f = f(x)$ defined for $x \in \mathbb{R}$ such that

$$\lim_{|x|\to\infty} f(x) = 0$$

and represent the error $\mathcal{E}^n$ corresponding to an interpolation by sinc functions with an increasing number of nodes.

Hint: For the computation, use the same program used in Example 2.4.2.

PROBLEM 2.10

Consider Problem 2.9 and develop the same analysis in the case of interpolation by Lagrange polynomials for the function $f(z(x))$ obtained using the change of variable defined in Remark 2.2.1.

Hint: Use again the program applied in Example 2.4.2.

3

Nonlinear Initial Value Problems in Unbounded Domains

3.1 Introduction

This chapter deals with the computational solution of nonlinear initial value problems described by partial differential equations when the space variable is defined over unbounded domains in one space dimension. The technical application of the solution method is developed for the classical Cauchy problem.

The method is based on the application of the collocation and interpolation technique described in Chapter 2. It generally uses interpolations by sinc functions which appear to be particularly well suited to approximate functions in unbounded domains with data decaying to zero at infinity. The applications refer to various models of solitary wave propagation introduced in Chapter 1. These numerical experiments are finalized to optimize the selection of the parameters for application of the algorithms. The solutions of the problems take advantage of the scientific programs reported in the Appendix, Sections A.3–A.8.

The chapter is developed through six more sections.

• Section 3.2 gives a detailed *description of the solution method* referred to first- and second-order equations. Subsequently, it is shown how the method can be generalized to the solution of higher-order equations. This generalization is also proposed in view of the applications later in the chapter that refer to higher-order equations.

• Section 3.3 deals with the *the solution of the third-order Korteweg–de Vries model.* The computational solutions are compared with some analytic results which can be obtained for some particular applications. These comparisons contribute to the proper selection of the number of collocation points.

• Section 3.4 develops a *similar application of the methods to the fifth-order Korteweg–de Vries model*, which is again focused on the selection of the number of collocation points.

• Section 3.5 reports some *additional computational results related to various wave models* reported in Chapter 1 with the aim of visualizing, by computational simulations, some interesting phenomena described by the models.

• Section 3.6 offers a brief introduction to *solution methods for ordinary differential equations*. In fact, the application of the method reduces mathematical problems for partial differential equations to the solution of an approximating system with finite degrees of freedom. Its solution is based on known techniques for ordinary differential equations, see Marasco and Romano (2001), which are summarized with the aim of making this chapter self-contained.

• Section 3.7 finally proposes *various problems* for the reader to use as tools for practicing the application of the method.

3.2 On the Solution of Initial Value Problems

This section deals with the solution of initial value problems for partial differential equations in an unbounded domain, with initial conditions decaying to zero at infinity. The application of the method is first referred to a general class of equations with second-order space derivatives, and then specialized to various types of models.

Let us consider the following class of second-order partial differential equations:

$$\frac{\partial u}{\partial t} = \eta(t,x,u)\frac{\partial u}{\partial x} + \mu\left(t,x,u,\frac{\partial u}{\partial x}\right)\frac{\partial^2 u}{\partial x^2} + \varepsilon f\left(t,x,u,\frac{\partial u}{\partial x}\right), \tag{3.2.1}$$

where the dimensionless dependent variable

$$u = u(t,x) : \quad [0,1] \times \mathbb{R} \to [0,1] \tag{3.2.2}$$

describes, in the mathematical model, the state of a real physical system, and where η, μ, and f are assumed to be given functions of their arguments. The perturbation parameter ε is not necessarily small. The time variable is defined over the interval $[0,1]$ after normalization with respect to a suitable critical computational time T. Simulations for time $t > 1$ must be interpreted as units of T.

The above class of equations, as we have seen from the various models described in Chapter 1, already includes a large variety of interesting models of applied sciences. The preceding particularization is proposed for tutorial aims; relatively more general cases can easily be handled by the reader after the technical contents proposed in what follows.

Referring to the above class of equations, the following initial value problem is stated.

Problem (Cauchy) 3.2.1. *Consider the initial value problem for the class of models described by Eq. (3.2.1) with given initial condition*

$$u(0,x) = \varphi(x)\,, \qquad \forall\, x \in \mathbb{R}\,, \tag{3.2.3}$$

where φ is a given smooth function of the space variable x, such that

$$\varphi(x) \to 0 \quad as \quad |x| \to \infty\,. \tag{3.2.4}$$

The application of the generalized collocation method for the solution of ***Problem 3.2.1*** takes advantage of the sinc interpolation method reported in Chapter 2, Section 2.2. The use of Lagrange interpolation is however technically possible as will be discussed later. The application of the method can be summarized as follows.

1. The space variable is discretized into a suitable set of equally spaced collocation points $I_x = \{x_{-n}, \ldots, x_n\}$, where

$$x_i = ih \qquad \text{for any relative integer} \quad i\,.$$

2. The dependent variable $u = u(t,x)$ is interpolated and approximated by the values $u_i(t) = u(t,x_i)$ as follows:

$$u(t,x) \cong u^n(t,x) = \sum_{i=-n}^{n} S_i(x) u_i(t)\,, \tag{3.2.5}$$

where the expression of the sinc function $Si(x) = S_i(x,h)$ is given by Eq. (2.2.18).

3. The above interpolation is used to approximate the partial derivatives of the variable u in the nodal points

$$\frac{\partial^r u}{\partial x^r}(t;x_i) \cong \sum_{j=-n}^{n} a_{ij}^{(r)}(n) u_j(t)\,, \qquad r \geq 1\,, \tag{3.2.6}$$

where

$$a_{ij}^{(r)}(n) = \frac{d^r S_j}{dx^r}(x_i). \tag{3.2.7}$$

The technical expressions of the above derivatives have been reported in Eqs. (2.2.21).

4. The initial value problem, for partial differential equations, is transformed into an initial value problem for ordinary differential equations describing the evolution of the values $u_i(t)$ of u in the nodes, with the additional assumption that $u_i = 0$ for $|i| > n$.
5. The solution of the initial value problem is then obtained solving the initial value problem for ordinary differential equations and interpolating the solution by the method used in Step 2. Specifically, the above procedure yields a system of ordinary differential equations which defines the time evolution of the values of the variable u in the nodal points

$$\frac{du_i}{dt} = \eta(t, x_i, u_i) \sum_{j=-n}^{n} a_{ji}^{(1)} u_j + \mu\left(t, x_i, u_i, \sum_{j=-n}^{n} a_{ji}^{(1)} u_j\right) \sum_{j=-n}^{n} a_{ji}^{(2)} u_j(s)$$
$$+ \varepsilon f\left(t, x_i, u_i, \sum_{j=-n}^{n} a_{ji}^{(1)} u_j\right), \tag{3.2.8}$$

with initial conditions

$$u_{i0} = u_i(t = 0) = \varphi(x_i), \tag{3.2.9}$$

for $i = -n, \ldots, n$, where we used the shorthand notation

$$a_{ij}^{(r)} = a_{ij}^{(r)}(n).$$

The system can be solved using standard techniques for ordinary differential equations, see Chapter 2 of Bellomo and Preziosi (1996); also additional information is given in the last section. The solution of (3.2.8)–(3.2.9) linked to the interpolation (3.2.5) provides the solution to Problem 3.2.1. The software ***Mathematica***® ensures the selection of the integration algorithms and of the time step for the integration.

The above method can be straightforwardly generalized to systems of partial differential equations, where the dependent variable is a vector, say **u**. In this case, the solution technique leads to a system of equations for each component of the vector variable **u**.

The generalization of the method to models with higher-order derivatives is also immediate, one simply has to use the approximation of higher-order space derivatives using formula (2.2.21). All examples dealt with in the next section refer to higher-order equations with reference to the models reported in Chapter 1.

Similar procedures can be developed for problems in the half-space, say $x \in [0, \infty)$, which need the boundary conditions at $x = 0$ and as $x \to \infty$, respectively:

$$u(t, 0) = \alpha(t)\,, \tag{3.2.10}$$

and

$$\lim_{x \to \infty} u(t, x) = 0\,. \tag{3.2.11}$$

In this case the interpolation can be applied in the half-space setting with $u_j = 0$ for any $j < 0$.

The use of Lagrange interpolations is technically possible. In such a case, it is convenient to introduce a change of variable to compact the real line into the space interval $[0, 1]$:

$$z = \frac{e^x}{1 + e^x}\,, \qquad x = \log \frac{z}{1 - z} = \psi(z)\,, \tag{3.2.12}$$

which implies

$$x \in [0, \infty) \Rightarrow z \in [0, 1)\,. \tag{3.2.13}$$

Therefore, Eq. (3.2.1) can then be rewritten as follows:

$$\begin{aligned} \frac{\partial u}{\partial t} &= \frac{1}{2}(1 - z^2)\eta(t, \psi(z))\frac{\partial u}{\partial z} \\ &+ \mu(t, \psi(z))\left[\frac{1}{4}(1 - z^2)^2\frac{\partial^2 u}{\partial z^2} - \frac{1}{2}z(1 - z^2)\frac{\partial u}{\partial z}\right] \\ &+ \varepsilon f\left(t, \psi(z), u, \frac{1}{2}(1 - z^2)\frac{\partial u}{\partial z}\right)\,. \end{aligned} \tag{3.2.14}$$

The application of the method generates a system of ordinary differential equations for $i = 1, \ldots, n$ such that $u_1 = 0\,, \quad u_n = 0$. The use of sinc functions has the advantage that equally spaced nodes for the variable z generate a discretization of the x variable such that the sequence $(x_i - x_{i-1})_i$ is nondecreasing, that is, computationally consistent when the data decay to zero at infinity.

3.3 On the Solution of the Third-Order KdV Model

This section deals with the computational solution of the initial value problem for the third-order Korteweg–de Vries model, which reads (see Chapter 1, Example 1.2.2.b) as follows:

$$\frac{\partial u}{\partial t} + u^m \frac{\partial u}{\partial x} + \mu \frac{\partial^3 u}{\partial x^3} = 0 . \tag{3.3.1}$$

Computations are developed for the intial value problem with initial condition

$$u_0(x) = [A \operatorname{sech}^2(kx - x_0)]^{1/m} , \tag{3.3.2}$$

where

$$A = 2\mu k^2 \frac{(m+1)(m+2)}{m^2} . \tag{3.3.3}$$

The problem (3.3.1)–(3.3.2) admits an analytic solution, which is called a ***solitary wave*** solution, of the type

$$u(t,x) = [A \operatorname{sech}^2(kx - \omega t - x_0)]^{1/m} , \tag{3.3.4}$$

where $\omega = 4\mu k^3/m^2$. Calculations have been developed for

$$m = 1 , \quad \mu = 1 , \quad k = 0.3 , \quad x_0 = 0 .$$

The first five invariants, see Whitham (1974) and Johnson (1997), related to the above solutions are

$$I_1 = \int_{-\infty}^{+\infty} u \, dx ,$$

$$I_2 = \int_{-\infty}^{+\infty} \frac{1}{2} u^2 \, dx ,$$

$$I_3 = \int_{-\infty}^{+\infty} \left(u^3 + \frac{1}{2} u_x^2 \right) \, dx ,$$

$$I_4 = \int_{-\infty}^{+\infty} \left(5u^4 + 10 u u_x^2 + u_{xx}^2 \right) \, dx ,$$

and

$$I_5 = \int_{-\infty}^{+\infty} \left(21u^5 + 105u^2 u_x^2 + 21uu_{xx}^2 + u_{xxx}^2\right) dx\,,$$

where subscripts to the variable u have been used to denote partial derivatives.

Computations have been referred to the space and time intervals, respectively: $x \in [-30, 30]$ and $t \in [0, 40]$ which are scaled to the interval $[0, 1]$ for both variables, as visualized in Figure 3.3.1. Computations have been performed with the help of the program ***KdVIII***, see the Appendix, Section A.7.

The system of ordinary differential equations corresponding to the above model is obtained by an immediate technical generalization of the method described in Section 3.2. The result is the following:

$$\frac{du_i}{dt} = -u_i \sum_{j=1}^{n} a_{ji}^{(1)} u_j - \sum_{j=1}^{n} a_{ji}^{(3)} u_j\,, \tag{3.3.5}$$

where, according to the analysis of Chapter 2, Eq. (2.2.21),

$$a_{ji}^{(1)} = \frac{(-1)^{i-j}}{h(i-j)}\,, \qquad a_{ii}^{(1)} = 0\,, \tag{3.3.6}$$

and

$$a_{ji}^{(3)} = \frac{(-1)^{i-h}}{h^3}\left(\frac{\pi^2}{i-j} - \frac{6}{(i-j)^3}\right), \qquad a_{ii}^{(3)} = 0\,. \tag{3.3.7}$$

The system of ordinary differential equations obtained inserting the boundary conditions is as follows:

$$\begin{cases} \dfrac{du_1}{dt} = -9.3312\,\mathrm{sech}^2(9 + 4.32t)\tanh(9 + 4.32t)\,, \\ \vdots \\ \dfrac{du_i}{dt} = -u_i \displaystyle\sum_{j=1}^{n} a_{ji}^{(1)} u_j - \sum_{j=1}^{n} a_{ji}^{(3)} u_j\,, \\ \vdots \\ \dfrac{du_n}{dt} = 9.3312\,\mathrm{sech}^2(9 - 4.32t)\tanh(9 - 4.32t)\,, \end{cases} \tag{3.3.8}$$

to be solved with initial conditions

$$\begin{cases} u_1(0) = 6.57935\,10^{-8}\,, \\ \quad\vdots \\ u_i(0) = 1.08\mathrm{sech}^2(9 - 18\,x_i)\,, \\ \quad\vdots \\ u_n(0) = 6.57935\,10^{-8}\,. \end{cases} \tag{3.3.9}$$

Simulations have been developed with 51 nodes, with the aim of obtaining accuracy on the third digit. Comparisons are reported in Table 3.3.1 which shows a very good agreement between the analytical and computational results. The solution is represented in Figure 3.3.1. The result of the simulation is very regular and smooth, despite the small number of nodes. The soliton translates without deformation, while no irregularities are visible in the flat portions of the domain. The second column of Table 3.3.1 reports the analytical value (a.v.) of the invariants which is compared with the numerical ones. The experiment shows that increasing the number of nodes does not substantially improve the accuracy of the solution. This is due to the fact that although increasing n improves the accuracy of interpolation, this improvement is counterbalanced by the error induced by increasing the number of ordinary differential equations to integrate.

Table 3.3.1

	a.v.	$t = 0$	$t = 0.25$	$t = 0.50$	$t = 0.75$	$t = 1$
I_1	7.200	7.200	7.200	7.200	7.199	7.198
I_2	2.592	2.592	2.592	2.592	2.592	2.592
I_3	4.666	4.666	4.666	4.666	4.665	4.666
I_4	23.131	23.131	23.194	23.129	23.127	23.130
I_5	102.792	102.793	102.778	102.776	102.757	102.784

Additional calculations can be developed in the case of two solitary waves for the same model, referred to the solution

$$u(t,x) = \sum_{i=1}^{2} \left[A_i\,\mathrm{sech}^2\left(k_i x - \omega_i t - x_i\right)\right]^{1/m}\,, \tag{3.3.10}$$

where

$$A_i = 2(m+1)(m+2)m^{-2}\mu k_i^2\,,$$

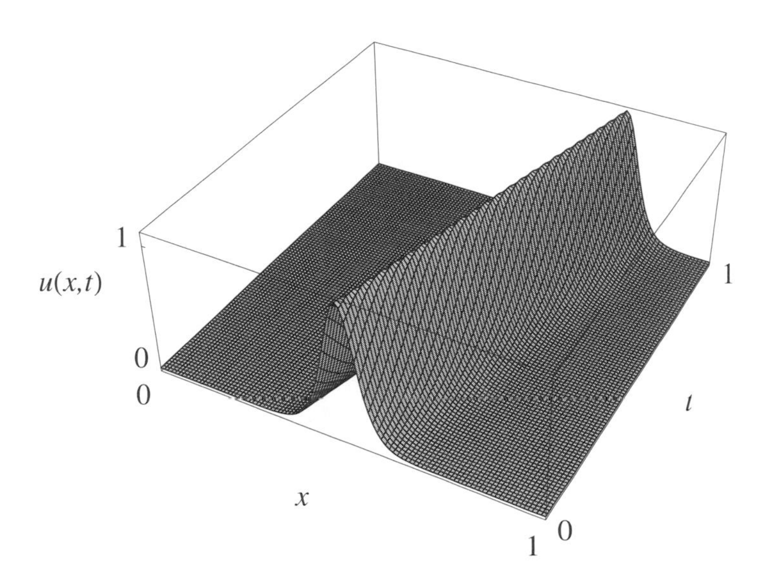

Figure 3.3.1 - Simulations (with 51 nodes) of the third-order KdV model.

and

$$\omega_i = 4\mu k_i^3 m^{-2}, \tag{3.3.11}$$

where k_i, x_i are given real numbers.

Table 3.3.2

	$t = 0$	$t = 0.25$	$t = 0.50$	$t = 0.75$	$t = 1$
I_1	15.3905	15.3923	15.3986	15.3760	15.3780
I_2	3.17387	3.17389	3.17389	3.17386	3.17408

Computations have been developed for the following values of the variables: $x \in [-70, 70]$, $t \in [0, 360]$, $\mu = 1$, $k_1 = 0.3$, $k_2 = 0.2$, $x_1 = -2$, and $x_2 = 3$. Calculations corresponding to the first two invariants are reported in Table 3.3.2 while the solution, for 81 nodes, is represented in Figure 3.3.2, again for the space variable scaled in the interval $[0, 1]$. Note that no analytical value of the first two invariants is known. Also in this case the result is remarkable, considering that the interaction between the solitons is simulated very carefully: both the peak values and the different propagation velocities appear clearly in Figure 3.3.2.

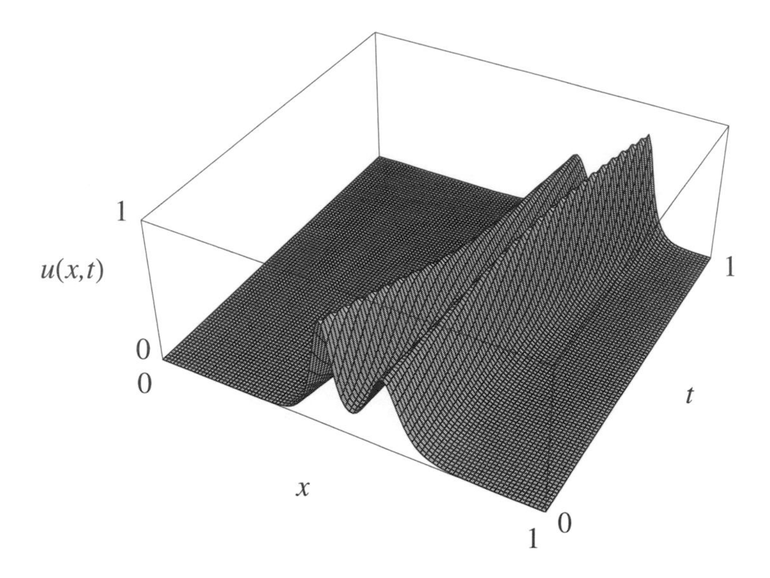

Figure 3.3.2 - Simulations (with 81 nodes) of two waves.

3.4 On the Solution of the Fifth-Order KdV Model

This section develops an analysis analogous to the one of Section 3.3 with reference now to the initial value problem for the fifth-order Korteweg–de Vries equation reported in Example 1.2.2.b, which reads as follows:

$$\frac{\partial u}{\partial t} + u\frac{\partial u}{\partial x} + \frac{\partial^3 u}{\partial x^3} - \frac{\partial^5 u}{\partial x^5} = 0\,. \tag{3.4.1}$$

The discrete system is

$$\frac{du_i}{dt} = -u_i \sum_{j=-n}^{n} a_{ji}^{(1)} u_j - \sum_{j=1}^{n} a_{ji}^{(3)} u_j + \sum_{j=1}^{n} a_{ji}^{(5)} u_j\,, \tag{3.4.2}$$

where, according to the calculations of Section 2.2 of Chapter 2,

$$a_{ji}^{(5)} = \frac{(-1)^{i-j}}{h^5 (i-j)^5} \sum_{k=0}^{5} (-1)^{k+1} \frac{5!}{(2k+1)!} \pi^{2k} (i-j)^{2k}\,, \quad a_{ii}^{(5)} = 0\,. \tag{3.4.3}$$

Also for this fifth-order equation, some analytical solutions of (3.4.1) are known, see Kaya (2004). In particular, if the following initial condition is given:

$$u(t=0,x)=u_0(x)=\frac{105}{169}\operatorname{sech}^4\left[\frac{x-x_0}{2\sqrt{13}}\right], \tag{3.4.4}$$

where x_0 is a free parameter, the solitary wave solution is as follows:

$$u(t,x)=\frac{105}{169}\operatorname{sech}^4\left[\frac{1}{2\sqrt{13}}\left(x-x_0-\frac{36}{169}t\right)\right]. \tag{3.4.5}$$

Application of the method to the third-order KdV model yields

$$\begin{cases} \dfrac{du_1}{dt} = -7.34136\operatorname{sech}^4(4.16025+2.95402t)\tanh(4.16025+2.95402t)\,,\\ \quad\vdots \\ \dfrac{du_i}{dt} = -u_i\displaystyle\sum_{j=1}^{n} a_{ji}^{(1)}u_j - \sum_{j=1}^{n} a_{ji}^{(3)}u_j + \sum_{j=1}^{n} a_{ji}^{(5)}u_j\,,\\ \quad\vdots \\ \dfrac{du_n}{dt} = 7.34136\operatorname{sech}^4(8.3205-2.95402t)\tanh(8.3205-2.95402t)\,, \end{cases} \tag{3.4.6}$$

with initial conditions

$$\begin{cases} u_1(0) = 5.88712\,10^{-7}\,,\\ \quad\vdots \\ u_i(0) = 0.621302\operatorname{sech}^4(12.48075\,x_i-4.16025)\,,\\ \quad\vdots \\ u_n(0) = 5.88712\,10^{-7}\,. \end{cases} \tag{3.4.7}$$

Simulations obtained using 81 nodes are reported in Figure 3.4.1. It can be observed that the simulation is very stable and no spurious instabilities are present. The program ***KdVV*** used for the simulations is reported in Section A.8 of the Appendix.

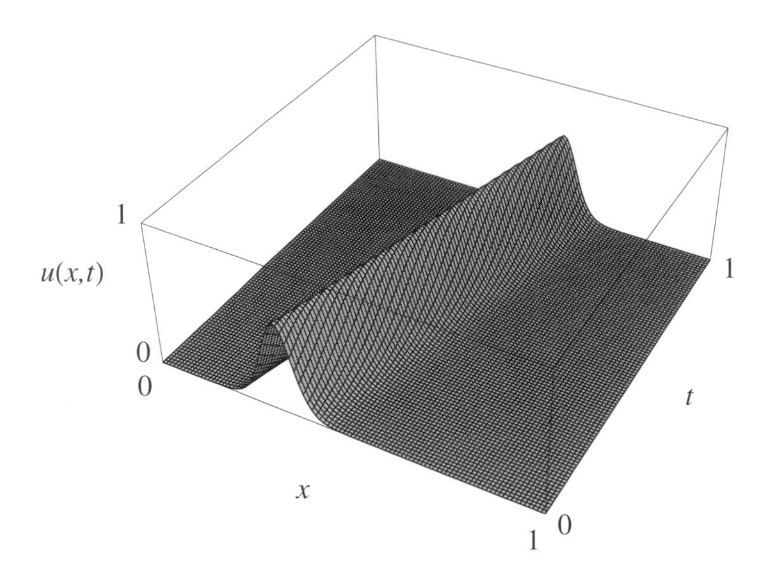

Figure 3.4.1 - Simulations (with 81 nodes) of the fifth-order KdV model.

3.5 Additional Applications and Discussion

The applications in Sections 3.3 and 3.4 have shown how the generalized collocation method can be applied to the solution of nonlinear problems in one space dimension in unbounded domains. Simulations have been developed by using sinc interpolation which shows the technical advantage of describing naturally the trend to zero at infinity. Various computational experiments suggest the use of this interpolation method although Lagrange polynomials can be technically used. On the other hand, Lagrange interpolation appears to be relatively more efficient for boundary value problems in bounded domains, as is will be shown in Chapter 4.

The two applications in the previous sections refer precisely to the models reported in Example 1.2.2 of Chapter 1, by Eqs. (1.2.9) and (1.2.11). These models are known as the third-order and fifth-order Korteweg–de Vries equations. The selection of the applications is motivated by the possibility of exploiting some analytic solutions, for particular initial conditions, to compare them to computational results.

Computations developed and visualized in the previous sections have clearly shown how satisfactory approximations of the analytic solution can be obtained by a proper selection of the number of nodes. Indeed, the

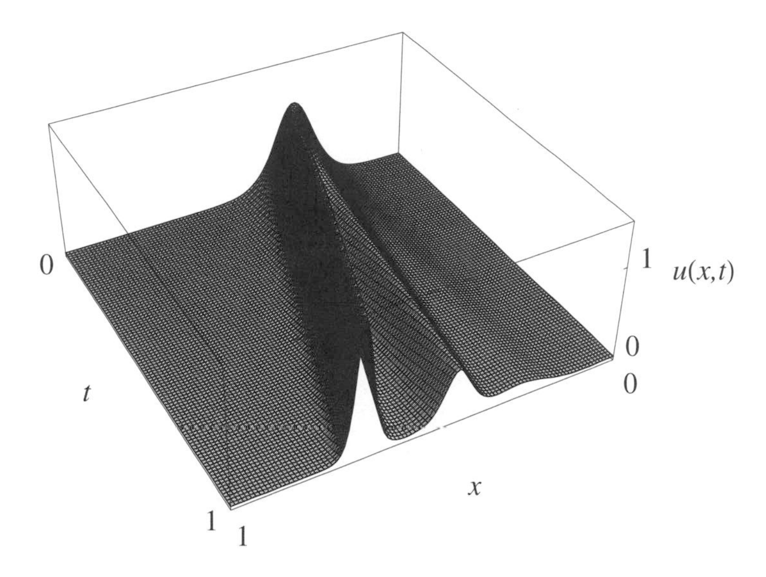

Figure 3.5.1 - Solution of the dispersive KdV model.

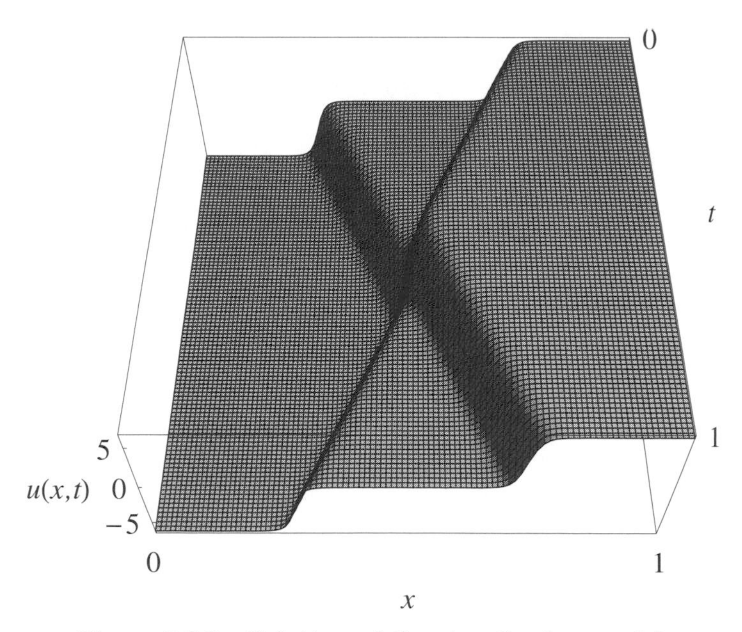

Figure 3.5.2 - Solution of the sine-Gordon model.

optimization of the number of nodes is a delicate topic which has not yet been exhaustively dealt with in this chapter. On the other hand, additional analyses will be proposed in the next chapters referring to various problems.

Therefore, leaving the above problem technically open, some additional simulations are proposed in this section without reporting the scientific programs to obtain them; we assume that the reader already has the necessary information and skills to develop programs ***KdVIII*** and ***KdVV*** to obtain the simulations reported in what follows.

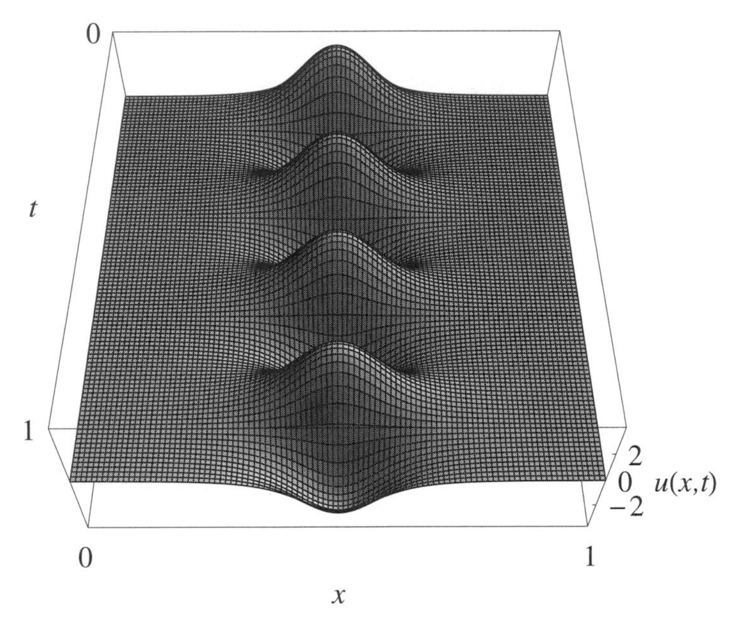

Figure 3.5.3 - Visualization of the breather solution.

• Figure 3.5.1 shows the solution of the dispersive KdV model, described by Eq. (1.2.12) of Chapter 1. In this case the effect of dispersion is well captured by the simulation. The correct number of solitons emerges and their dynamics are described without instability.

• Figure 3.5.2 visualizes the solution of the sine-Gordon model, reported in Example 1.2.3 of Chapter 1; simulations are developed for $c_1 = c_2 = 0$ and $m = 2$.

• Figure 3.5.3 visualizes the breather solution, reported in Eq. (1.2.16), and illustrates satisfactory results obtained by interpolation and collocation using Lagrange polynomials.

3.6 On the Solution of Ordinary Differential Equations

Generalized collocation methods reduce the solution of systems with infinite degrees of freedom to the solution of systems with finite degrees of freedom. Therefore, it is useful to summarize the main lines of the qualitative analysis and solution techniques for ordinary differential equations, so that the overall presentation of the method is self-consistent. The reader is referred to the pertinent literature, e.g., Bellomo and Preziosi (1996), Mickens (1996), Sewell (1998), and Marasco and Romano (2001), for a detailed overlook of this topic.

The application of collocation methods method leads to ***well-formulated problems*** for ordinary differential equations. The system is, in fact, made of a number of dependent variables equal to the number of linearly independent equations, and it is associated to the correct number (the same as the number of equations) of initial conditions. On the other hand, a correct statement does not imply that the mathematical problem is ***well posed***, i.e., that the solution exists, is unique, and depends continuously on the initial data and on the parameters of the model.

Let us consider the initial value problem

$$\begin{cases} \dfrac{d\mathbf{u}}{dt} = \mathbf{f}(t, \mathbf{u}) \\ \mathbf{u}_0 = \mathbf{u}(t_0) \end{cases}, \tag{3.6.1}$$

and let us introduce a norm in $\mathbb{R}^n$, which might be, for instance, the Euclidean norm

$$\|\mathbf{u}\| = \left(\sum_{i=1}^{n} u_i^2 \right)^{\frac{1}{2}} . \tag{3.6.2}$$

Several analytic properties of the solution are related to what is called the ***Lipschitz condition***. A vector function $\mathbf{f}(t, \mathbf{u})$ satisfies a ***Lipschitz condition*** in a region $\mathcal{D}$ of the $(t, \mathbf{u})$-space if there exists a constant L (called the ***Lipschitz constant***), such that, for any $(t, \mathbf{u})$ and $(t, \mathbf{v})$ in $\mathcal{D}$,

$$\big\|\mathbf{f}(t, \mathbf{u}) - \mathbf{f}(t, \mathbf{v})\big\| \leq L \, \|\mathbf{u} - \mathbf{v}\| , \tag{3.6.3}$$

where both terms on the left-hand side of (3.6.3) are referred to the same instant of time.

It can be proved that if $\mathbf{f}(t, \mathbf{u})$ is defined in a bounded, closed, and convex domain $\mathcal{D}$, and if the partial derivatives of $\mathbf{f}$ with respect to u_i exist with

$$\max_{i,j=1,\dots,n} \sup_{(t,\mathbf{u})\in\mathcal{D}} \left|\frac{\partial f_i}{\partial u_j}\right| \leq M\,, \tag{3.6.4}$$

then $\mathbf{f}$ satisfies a Lipschitz condition in convex domain $\mathcal{D}$ with Lipschitz constant equal to M.

Of course, if $\mathbf{f}(t,\mathbf{u})$ satisfies a Lipschitz condition in $\mathcal{D}$, $\mathbf{f}$ is a continuous function of $\mathbf{u}$ in $\mathcal{D}$ for each fixed t. On the other hand, if $\mathbf{f}(t,\mathbf{u})$ is differentiable with respect to $\mathbf{u}$, then it satisfies the Lipschitz condition, but the opposite is not true, as is shown by $|u|$ that satisfies the Lipschitz condition with constant $L = 1$ though it is not differentiable in $u = 0$.

Existence and uniqueness of the solution to problem (3.6.1) are assessed by the following classical theorems.

Theorem 3.6.1. ***(Existence).*** *If* $\mathbf{f}(t,\mathbf{u})$ *is continuous in the rectangle*

$$\mathcal{R} = \left\{(t,\mathbf{u}) : \|\mathbf{u}-\mathbf{u}_0\| \leq K\,,\ |t-t_0| \leq T\right\},$$

at least one solution to the initial value problem (3.6.1) exists there, and it is of class C^1 *for* $|t-t_0| \leq \widehat{T}$, *where*

$$\widehat{T} = \min\left\{T, \frac{K}{M}\right\} \qquad \text{and} \quad M = \max_{(t,\mathbf{u})\in\mathcal{R}} \left\|\mathbf{f}(t,\mathbf{u})\right\|.$$

Theorem 3.6.2. ***(Uniqueness).*** *If, in addition to the continuity condition of Theorem 3.6.1, the function* $\mathbf{f}$ *satisfies a Lipschitz condition in* $\mathcal{R}$, *then the solution* $\mathbf{u}(t)$ *to the initial value problem (3.6.1) is* ***unique***, *and*

$$\|\mathbf{u}(t) - \mathbf{u}_0\| \leq M\widehat{T}\,. \tag{3.6.5}$$

Remark 3.6.1. *The Lipschitz condition is not needed to ensure the existence of a solution of the initial value problem. On the other hand, it is essential in the uniqueness proof.*

The literature on ordinary differential equations reports various generalizations and improvements of the above theorems. The continuous dependence on the initial data and on $\mathbf{f}$ are stated by the following.

Theorem 3.6.3. ***(Continuous dependence on the initial data).*** *If* $\mathbf{f}$ *is continuous and satisfies the Lipschitz condition, then*

$$\left\|\widehat{\mathbf{u}}(t) - \widetilde{\mathbf{u}}(t)\right\| \leq \left\|\widehat{\mathbf{u}}_0 - \widetilde{\mathbf{u}}_0\right\| e^{L|t-t_0|}\,, \tag{3.6.6}$$

where $\widehat{\mathbf{u}}$ *and* $\widetilde{\mathbf{u}}$ *are the two solutions of the two initial value problem (3.6.1) with initial data* $\mathbf{u}(t_0) = \widehat{\mathbf{u}}_0$ *and* $\mathbf{u}(t_0) = \widetilde{\mathbf{u}}_0$, *respectively.*

Theorem 3.6.4. ***(Continuous dependence on* f*).*** *Let us consider for $t \in [t_0, t_0 + T]$ the initial value problems*

$$\begin{cases} \dfrac{d\mathbf{u}}{dt} = \mathbf{f}(t, \mathbf{u})\,, \\ \mathbf{u}(t_0) = \mathbf{u}_0\,, \end{cases} \qquad \text{and} \qquad \begin{cases} \dfrac{d\mathbf{u}}{dt} = \mathbf{f}^*(t, \mathbf{u})\,, \\ \mathbf{u}(t_0) = \mathbf{u}_0^*\,, \end{cases} \tag{3.6.7}$$

with $\mathbf{f}$ *and* $\mathbf{f}^*$ *defined and continuous in a common domain* $\mathcal{D}$*. If one of* $\mathbf{f}$ *or* $\mathbf{f}^*$ *satisfies a Lipschitz condition with constant* L *and if*

$$\big\|\mathbf{f}(t, \mathbf{u}) - \mathbf{f}^*(t, \mathbf{u})\big\| \leq \varepsilon\,, \qquad \forall\,(t, \mathbf{u}) \in \mathcal{D}\,, \tag{3.6.8}$$

then

$$\big\|\mathbf{u}(t) - \mathbf{u}^*(t)\big\| \leq \big\|\mathbf{u}_0 - \mathbf{u}_0^*\big\| e^{L|t-t_0|} + \frac{\varepsilon}{L}\Big[e^{L|t-t_0|} - 1\Big]\,, \tag{3.6.9}$$

where $\mathbf{u}(t)$ *and* $\mathbf{u}^*(t)$ *are the solutions of the two initial value problems defined in (3.6.7).*

Remark 3.6.2. *Theorems 3.6.1–3.6.4 give minimal conditions, while, the domain might be larger than the estimate given by the theorems, possibly extending to all $t \geq t_0$. Then one needs criteria to determine the largest possible domain of existence. This type of analysis is often based on a deep understanding of the qualitative properties of the system, by a qualitative analysis of energy conservation, that may imply that the solution is bounded.*

Computational schemes can be applied to problems which are well posed. These schemes are all based on the discretization of the time variable $t \in [t_0, T]$ into a suitable set of points:

$$I_t = \{t_0, \ldots, t_i, \ldots, t_n = T\}\,,$$

and by approximating the solution in the points of the discretization by suitable algorithms. The application needs to take into account ***truncation error***, ***accuracy***, and ***stability*** of the algorithms, according to the following definitions.

Definition 3.6.1. ***(Truncation error).*** *The truncation error $T_{\rm err}$ is defined as the norm of the difference between the solution to the differential equation and the numerical solution divided by the time step used in the numerical scheme.*

Definition 3.6.2. ***(Accuracy).*** *If the truncation error goes like h^p, the method is called of the pth order. This is what is generally meant by* ***accuracy of the method.***

Definition 3.6.3. ***(Absolute stability and stability region).*** *A numerical scheme is absolutely stable in a point ah of the complex plane if a sequence $\{u_i\}$ generated by the method applied to the linearized equation with time step $\Delta t = h$ is bounded for $i \to +\infty$. The* ***stability region*** *is the set of points ah in the complex plane for which the method is absolutely stable.*

It can be proved that each scheme is characterized by a stability region, which can be used to determine a condition on the time step to be used in the integration. The procedure to be used is the following:

Step 1. Starting from a nonlinear model, consider its linearized form

$$\frac{d\mathbf{u}}{dt} = \mathbf{A}\mathbf{u}\,, \tag{3.6.10}$$

obtained for instance by linearization about the initial condition;

Step 2. Compute the eigenvalues $\lambda_1, \ldots, \lambda_n$ of $\mathbf{A}$, called the spectrum of the linearized model;

Step 3. Choose, if possible, a not too restrictive time step h such that $h\lambda_1$, ..., $h\lambda_n$ all belong to the stability region.

Then the numerical errors are not amplified as the integration is continued. If, instead, there are some eigenvalues which remain outside the stability region, then one must be aware that the numerical errors grow exponentially in time. The literature on computational methods reports a large variety of algorithms to obtain simulations of finite models.

Consider, as an example of the algorithms, the ***Runge–Kutta methods*** which are very popular for their adaptability and versatility and work quite well for all nonstiff problems. They are obtained by evaluating the function f at different values of t and u, and by suitably combining these values. In this way it is possible to obtain methods of any order of accuracy. For instance, for second-order accuracy:

$$\text{2nd order}: \quad \mathbf{u}_{i+1} = \mathbf{u}_i + \frac{h}{2}\Big(\mathbf{K}_1 + \mathbf{K}_2\Big)\,, \tag{3.6.11}$$

where

$$\mathbf{K}_1 = \mathbf{f}(t_i, \mathbf{u}_i)\,, \qquad \mathbf{K}_2 = \mathbf{f}(t_i + h, \mathbf{u}_i + h\mathbf{K}_1)\,. \tag{3.6.12}$$

For third-order accuracy:

$$\text{3rd order :} \quad \mathbf{u}_{i+1} = \mathbf{u}_i + \frac{h}{6}\Big(\mathbf{K}_1 + 4\mathbf{K}_2 + \mathbf{K}_3\Big)\,, \tag{3.6.13}$$

where

$$\begin{aligned} \mathbf{K}_1 &= \mathbf{f}(t_i, \mathbf{u}_i)\,, \\ \mathbf{K}_2 &= \mathbf{f}\left(t_i + \frac{h}{2}, \mathbf{u}_i + \frac{h}{2}\mathbf{K}_1\right)\,, \\ \mathbf{K}_3 &= \mathbf{f}\big(t_i + h, \mathbf{u}_i + h(2\mathbf{K}_2 - \mathbf{K}_1)\big)\,. \end{aligned} \tag{3.6.14}$$

Finally for fourth-order accuracy:

$$\text{4th order :} \quad \mathbf{u}_{i+1} = \mathbf{u}_i + \frac{h}{6}\Big(\mathbf{K}_1 + 2\mathbf{K}_2 + 2\mathbf{K}_3 + \mathbf{K}_4\Big)\,, \tag{3.6.15}$$

where

$$\begin{aligned} \mathbf{K}_1 &= \mathbf{f}(t_i, \mathbf{u}_i)\,, \\ \mathbf{K}_2 &= \mathbf{f}\left(t_i + \frac{h}{2}, \mathbf{u}_i + \frac{h}{2}\mathbf{K}_1\right)\,, \\ \mathbf{K}_3 &= \mathbf{f}\left(t_i + \frac{h}{2}, \mathbf{u}_i + \frac{h}{2}\mathbf{K}_2\right)\,, \\ \mathbf{K}_4 &= \mathbf{f}(t_i + h, \mathbf{u}_i + h\mathbf{K}_3)\,. \end{aligned} \tag{3.6.16}$$

3.7 Problems

PROBLEM 3.1
Solve the problem in Section 3.3 (using the program ***KdVIII***) for a variable number of nodes thus visualizing the influence of the number of nodes over the accuracy of the solution.

PROBLEM 3.2
Develop the same analysis of Problem 3.1 with reference to the problem in Section 3.4.

PROBLEM 3.3
Apply the change of variable (3.2.13) to compact the real line into the interval $[0, 1]$ for the problem in Section 3.3 and solve the problem using Lagrange-type interpolation.

PROBLEM 3.4
Apply the same analysis of Problem 3.3 to the solution of the problem in Section 3.4.

PROBLEM 3.5
Develop a comparison between the computational solution of the initial value problem in this chapter using Lagrange interpolation compared to sinc interpolation.

Hint: Follow the same procedure applied in Problem 3.1.

PROBLEM 3.6
For the problems in Sections 3.3 and 3.4, compare the spatial derivatives with the analytical ones.

PROBLEM 3.7
Analyze the solution of the problem considered in Section 3.3 using the same number of nodes, but in a space interval first smaller and then larger than the one used in that application.

PROBLEM 3.8
Consider the problem in Section 3.3 in a larger space interval and identify the number of nodes necessary to obtain the same accuracy than the one obtained in Section 3.3.

PROBLEM 3.9
Apply the same analysis of Problem 3.7 to the model in Section 3.4.

PROBLEM 3.10
Develop an analysis analogous to that of Problems 3.7 and 3.8 to the models considered in Section 3.5.

4

Nonlinear Initial-Boundary Value Problems in One Space Dimension

4.1 Introduction

This chapter deals with the application of the generalized collocation method to the *solution of nonlinear initial-boundary value problems for partial differential equations in one space dimension.*

The technical difference with respect to Chapter 3, which discussed the Cauchy problem in unbounded domains, is the implementation of the ***boundary conditions***. It turns out to be quite a technical problem in the case of Dirichlet or Neumann boundary conditions, and nontrivial difficulties have to be dealt with in the case of nonlinear boundary conditions, as we shall see later in Chapter 6.

The application of the method makes use of the Lagrange polynomials, which allow a flexible implementation of the boundary conditions. The use of sinc functions for interpolation over bounded domains appears to be generally less efficient.

The applications proposed in this chapter refer to the traffic flow model described in Chapter 1, see Example 1.2.5, and to the nonlinear convection diffusion model, described in Example 1.3.1. The contents are organized through six more sections.

- Section 4.2 deals with *methodological topics* and specifically with the description of the application of the method to the classical Dirichlet, Neumann, and Robin problems: namely problems with linear boundary conditions. The application refers first to second-order scalar equations, and then technical generalizations to higher-order partial differential equations or systems of equations are discussed.

This section also shows how the method can be applied to the solution of problems in the half-space with boundary conditions decaying to zero at infinity.

• Section 4.3 shows how the method described in the preceding section can be applied to the solution of *problems with Dirichlet boundary conditions for traffic flow models.* Various types of boundary conditions, corresponding to real traffic flow conditions, are used for the simulations.

• Section 4.4 develops a *similar analysis for Neumann and Robin boundary conditions.* The application refers to the nonlinear convection diffusion equation.

• Section 4.5 shows how *problems with analytic solutions* can be obtained for models with source terms. The knowledge of analytic solutions allows their comparison with those obtained by the computational method.

• Section 4.6 develops a *technical analysis* concerning the *error estimates* and the selection of the *optimal number of collocation points.*

• The last section proposes, as in the previous chapters, *various problems* suitable for practicing the application of the method.

4.2 Problems with Linear Boundary Conditions

We apply the method in this section, with tutorial aims, to the following general class of scalar second-order partial differential equations:

$$\frac{\partial u}{\partial t} = f\left(t, x, u, \frac{\partial u}{\partial x}, \frac{\partial^2 u}{\partial x^2}\right), \tag{4.2.1}$$

where the dimensionless dependent variable

$$u = u(t,x) : \quad [0,1] \times [0,1] \to [0,1] \tag{4.2.2}$$

describes, in the mathematical model, the state of a real physical system, and f is assumed to be a given function of its arguments. Technical generalizations are dealt with later in this chapter with reference to systems of equations and equations with a higher order of derivatives.

The mathematical statement of the initial-boundary value problem was already given in Chapter 1. This section reports the statement of two-point Dirichlet, Neumann, and mixed problems.

Let us consider the following initial-boundary value problems with linear boundary conditions.

Problem (Dirichlet, Neumann, Mixed) 4.2.1. *The initial-boundary value problem for Eq. (4.2.1) is stated with initial condition*

$$u(0,x) = \varphi(x)\,, \qquad \forall\, x \in [0,1]\,, \tag{4.2.3}$$

and with boundary conditions:

i) ***Dirichlet***

$$u(t,0) = \alpha(t) \quad \textit{and} \quad u(t,1) = \beta(t)\,, \qquad \forall\, t \in [0,1]\,; \tag{4.2.4}$$

ii) ***Neumann***

$$\frac{\partial u}{\partial x}(t,0) = \gamma(t) \quad \textit{and} \quad \frac{\partial u}{\partial x}(t,1) = \delta(t)\,, \qquad \forall\, t \in [0,1]\,; \tag{4.2.5}$$

iii) ***mixed***

$$u(t,0) = \alpha(t) \quad \textit{and} \quad \frac{\partial u}{\partial x}(t,1) = \delta(t)\,, \qquad \forall\, t \in [0,1]\,, \tag{4.2.6}$$

or

$$\frac{\partial u}{\partial x}(t,0) = \gamma(t) \quad \textit{and} \quad u(t,1) = \beta(t)\,, \qquad \forall\, t \in [0,1]\,, \tag{4.2.7}$$

where φ, is a given function of space, and α, β, γ, and δ are given smooth functions of time, consistent with the initial condition (4.2.3), i.e.,

$$\alpha(0) = \varphi(0) \qquad \textit{and} \qquad \beta(0) = \varphi(1),$$

and

$$\gamma(0) = \frac{d\varphi}{dx}(0) \quad \textit{and} \quad \delta(0) = \frac{d\varphi}{dx}(1)\,.$$

The application of the method to the solution of ***Problem 4.2.1*** can be regarded as a technical generalization of the method to solve the initial value problem in unbounded domains. The technical difference is that the space variable is now defined on the interval $[0,1]$ and that boundary conditions have to be implemented.

Lagrange interpolation corresponding to a Chebychev collocation will be used with specific reference to the formulae reported in Chapter 2, Section 2.2. The use of sinc interpolation is however technically possible, as will be discussed later. Based on the above considerations, the application of the method consists in the following steps.

1. The space variable is discretized into the collocation $I_x = \{x_i\}_{i=1,\dots,n}$ where $x_1 = 0$, $x_n = 1$. In general, a Chebychev localization is used.
2. The dependent variable $u = u(t,x)$ is interpolated and approximated through the values $u_i(t) = u(t,x_i)$ as follows:

$$u(t,x) \cong u^n(t,x) = \sum_{i=1}^{n} L_i(x) u_i(t)\,, \tag{4.2.8}$$

where the expression of the Lagrange polynomials $L_i(x)$ is given by Eq. (2.2.6).
3. The above interpolation is used to approximate the partial derivatives of the variable u in the nodal points. We recall the formula already given in Chapter 2:

$$\frac{\partial^r u}{\partial x^r}(t;x_i) \cong \sum_{h=1}^{n} a_{hi}^{(r)}(n) u_h(t)\,, \qquad r = 1,2\,, \tag{4.2.9}$$

where

$$a_{hi}^{(r)}(n) = \frac{d^r L_h}{dx^r}(x_i)\,,$$

with technical expressions reported in Chapter 2, Eqs. (2.2.9)–(2.2.11).
4. The initial-boundary value problem is transformed into an initial value problem for ordinary differential equations describing the evolution of the values $u_i(t)$ of u in the nodes. Boundary conditions are imposed in the nodal points on the boundaries x_1 and x_n of the domain of the space variable.
5. The solution of the initial-boundary value problem is obtained by solving the initial value problem for ordinary differential equations and interpolating the solution by (4.2.8).

The above procedure generates a system of ordinary differential equations which defines the time evolution of the values of the variable u in the nodal points, which can be formally written as follows:

$$\begin{cases} u_1 = \alpha(t)\,, \\ \quad \vdots \\ \dfrac{du_i}{dt} = f\left(t, x_i, u_i, \displaystyle\sum_{j=1}^{n} a_{ji}^{(1)} u_j, \sum_{j=1}^{n} a_{ji}^{(2)} u_j\right), \quad i = 2,\dots,n-1\,, \\ \quad \vdots \\ u_n = \beta(t)\,. \end{cases} \tag{4.2.10}$$

Remark 4.2.1. *If technically useful, the first and last equations can be written in differential form:*

$$\frac{du_1}{dt} = \frac{d\alpha}{dt} \quad \text{and} \quad \frac{du_n}{dt} = \frac{d\beta}{dt},$$

so that all equations (from $i = 1$ to $i = n$) are differential equations.

The above system (4.2.10), once linked to the initial conditions

$$\varphi_{0i} = \varphi_0(x_i), \qquad i = 1, \ldots, n, \tag{4.2.11}$$

can be solved by means of standard techniques for ordinary differential equations, so that the evolution in time of the variable u in the nodes $u_i(t)$ is computed. The interpolation (4.2.8) provides the continuous-in-space solution to Problem 4.2.1.

The application of the method to the solution of ***Problem 4.2.1*** (Neumann) can be developed analogously. The difference consists in a different way to deal with the first and last equations where Neumann boundary conditions have to be imposed. Consistency with the boundary conditions yields the following linear algebraic system:

$$\begin{cases} a_{11}u_1 + a_{n1}u_n = \gamma - \displaystyle\sum_{j=2}^{n-1} a_{j1}u_j\,, \\[2ex] a_{1n}u_1 + a_{nn}u_n = \delta - \displaystyle\sum_{j=2}^{n-1} a_{jn}u_j\,, \end{cases} \tag{4.2.12}$$

where all terms u_i, γ, and δ have to be regarded as functions of time, and where we used the shorthand notation $a_{ji} := a_{ji}^{(1)}$ for any $i, j \in \{1, \ldots, n\}$.

The algebraic solution of the above system with respect to u_1 and u_n yields the following formal expression:

$$\begin{aligned} u_1 &= u_1(\gamma, \delta, u_2, \ldots, u_{n-1})\,, \\ u_n &= u_n(\gamma, \delta, u_2, \ldots, u_{n-1})\,. \end{aligned} \tag{4.2.13}$$

Substituting u_1 and u_n into Eqs. (4.2.10) yields a self-consistent system of ordinary differential equations corresponding to the model with Neumann boundary conditions.

Also in this case, differential equations for the first and last equations can be used. In fact, system (4.2.12) can be derived with respect to time with the following result:

$$\begin{cases} a_{11}\dfrac{du_1}{dt} + a_{n1}\dfrac{du_n}{dt} = \gamma - \displaystyle\sum_{j=2}^{n-1} a_{j1}\dfrac{du_j}{dt}\,, \\ a_{1n}\dfrac{du_1}{dt} + a_{nn}\dfrac{du_n}{dt} = \delta - \displaystyle\sum_{j=2}^{n-1} a_{jn}\dfrac{du_j}{dt}\,. \end{cases}$$

The same algebraic separation generates a differential system corresponding to system (4.2.13) in terms of $\dfrac{du_i}{dt}$, for $i = 1, \ldots, n$.

Programming the solution of the Neumann problem simply requires the addition of a few strings with respect to the solution of the Dirichlet problem. The strings refer to the solution of the above algebraic problem. On the other hand, it is plain that the computational complexity is increased because the solution of the first and last equation can only be approximated, while in the former case it was exact, being precisely stated by the boundary conditions.

Relatively simpler is the case of mixed boundary conditions: Dirichlet in $x = 0$ and Neumann in $x = 1$, or vice versa. The corresponding algebraic system is as follows:

$$\begin{cases} u_1 = \alpha\,, \\ a_{1n}u_1 + a_{nn}u_n = \delta - \displaystyle\sum_{j=2}^{n-1} a_{jn}u_j\,, \end{cases} \tag{4.2.14}$$

or

$$\begin{cases} a_{11}u_1 + a_{n1}u_n = \gamma - \displaystyle\sum_{j=2}^{n-1} a_{j1}u_j\,, \\ u_n = \beta\,. \end{cases} \tag{4.2.15}$$

In the first case, Eq. $(4.2.14)_2$ can be used to obtain the second boundary condition:

$$u_n = u_n(\alpha, \delta, u_2, \ldots, u_{n-1})\,, \tag{4.2.16}$$

where $\alpha = \alpha(t)$ and $\delta = \delta(t)$ are given.

In the second case, Eq. $(4.2.15)_1$ can be used to obtain the first boundary condition:

$$u_1 = u_1(\beta, \gamma, u_2, \ldots, u_{n-1})\,, \tag{4.2.17}$$

where $\beta = \beta(t)$ and $\gamma = \gamma(t)$ are given.

The generalization to the case of the Robin problem is immediate. The problem is stated as a linear combination of the Dirichlet and Neumann problems in $x = 0$ and $x = 1$.

Problem (Robin) 4.2.2. *Consider the initial-boundary value problem for Eq. (4.2.1) with initial condition (4.2.3) and boundary conditions*

$$a(t) = c_1\,\alpha(t) + c_2\,\gamma(t)\,, \qquad \forall\, t \in [0,1]\,, \tag{4.2.18a}$$

and

$$b(t) = c_3\,\beta(t) + c_4\,\delta(t)\,, \qquad \forall\, t \in [0,1]\,, \tag{4.2.18b}$$

where c_i $(i = 1,2,3,4)$ are given constants, and $a = a(t)$ and $b = b(t)$ are given smooth functions of time, consistent, for $t = 0$, with the initial conditions.

The technical solution of such a problem follows the same lines we have seen above. In fact, recall that γ and δ can be approximated by the following Lagrange interpolation:

$$\gamma = a_{11}\alpha + \sum_{j=2}^{n-1} a_{j1}u_j + a_{n1}\beta\,, \tag{4.2.19}$$

and

$$\delta = a_{1n}\alpha + \sum_{j=2}^{n-1} a_{jn}u_j + a_{nn}\beta\,. \tag{4.2.20}$$

Inserting the above expression into (4.2.18) generates an algebraic system in the unknown α and β analogous to the one we have seen for the Neumann problem.

Remark 4.2.2. *The above described mathematical method has been technically referred to the class of models (4.2.1) for problems on a segment* $[0,1]$. *Note that the solution method of the Dirichlet problem can be dealt with using both Lagrange and sinc interpolations. However, recalling that in case of sinc interpolation one has $a_{ii} = 0$, then the local coordinates on the boundary do not contribute to the space derivative. Therefore Neumann boundary conditions cannot be straightforwardly implemented.*

Some technical generalizations are immediate. For instance, the above method can also be applied to the case of the Dirichlet Problem 4.2.1 in the half-space with initial conditions $u(0,x) = \varphi(x)$, $\varphi : \mathbb{R} \to \mathbb{R}_+$ and boundary conditions

$$u(0,x) = \alpha(t)\,, \qquad \lim_{x\to\infty} u(t,x) = 0\,. \tag{4.2.21}$$

The solution of this problem has been introduced in Chapter 3. It can be shown that the half-space can be compacted into the domain $[0, 1)$ by the change of variable

$$z = \frac{e^x - 1}{e^x + 1}, \qquad x = \log\frac{1+z}{1-z} = \psi(z), \tag{4.2.22}$$

such that $z \in [0, 1)$ yields $x \in [0, \infty)$. Consequently, Eq. (4.2.1) can be rewritten as follows:

$$\frac{\partial u}{\partial t} = f\left(t, \psi(z), \frac{1}{2}(1-z^2)\frac{\partial u}{\partial z}, \frac{1}{4}(1-z^2)^2\frac{\partial^2 u}{\partial z^2}\right). \tag{4.2.23}$$

The application of the method follows immediately by yielding a system of ordinary differential equations for $i = 1, \ldots, n$ such that $u_1(t) = \alpha(t)$ and $u_n = 0$.

Similarly, one can deal with systems of partial differential equations. This generalization, which is also immediate, refers to the case where the dependent variable is a vector, say $\mathbf{u}$. The solution technique leads to a system of equations for each component of $\mathbf{u}$. In particular, problems with second-order time derivatives can be reduced to a first-order system of equations in the augmented variable

$$\mathbf{v} = \{v_1, v_2\} = \left\{u, \frac{\partial u}{\partial t}\right\}. \tag{4.2.24}$$

For instance, the wave equation

$$\frac{\partial^2 u}{\partial t^2} = \frac{\partial^2 u}{\partial x^2} \tag{4.2.25}$$

can be written, by using (4.2.24), as a system of partial differential equations as follows:

$$\begin{cases} \dfrac{\partial v_1}{\partial t} = v_2\,, \\[2ex] \dfrac{\partial v_2}{\partial t} = \dfrac{\partial^2 v_1}{\partial x^2}\,. \end{cases} \tag{4.2.26}$$

Initial conditions have to be imposed both for v_1 and v_2 as both equations involve time derivatives, while boundary conditions have to be imposed for v_2 only, since the first equation does not involve space derivatives.

The same procedure can be applied to the general case of higher-order equations. The application of the methods is again developed through two steps. First the continuous system is discretized into a finite system of n

equations precisely as we have seen in Chapter 3. Then, boundary conditions are implemented in x_1 and x_n. Of course, systems with higher-order space derivatives need additional boundary conditions which contain also x_2 and/or x_{n-1}.

The examples proposed in the following section clarify this matter. The optimal selection of the number of nodes plays a central role in assessing the reliability of the method. This important topic is discussed later.

4.3 Traffic Flow Model with Dirichlet Boundary Conditions

The mathematical method described in the preceding section is here applied to the solution of the initial-boundary value problem, with Dirichlet boundary conditions, for the traffic flow model described in Example 1.2.5.

Let us consider the model stated by Eq. (1.2.28), which is written here again:

$$\frac{\partial u}{\partial t} = (2u-1)\frac{\partial u}{\partial x} + \eta u^2(1-u)\frac{\partial^2 u}{\partial x^2} + \eta u(2-3u)\left(\frac{\partial u}{\partial x}\right)^2, \qquad (4.3.1)$$

where u is the dimensionless density of the vehicles, while the dependent variables are the time t and the space coordinate $x \in [0,1]$. The parameter η refers to the sensitivity of the driver to the space gradients of the density. The following specific Dirichlet problem is dealt with.

Problem 4.3.1. *The initial-boundary value problem is stated by linking Eq. (4.3.1) to the following initial and boundary conditions:*

$$u_0(x) = u(t=0,x) = 0.2\,, \qquad (4.3.2)$$

and

$$\begin{cases} \alpha(t) = u(t,x=0) = 0.5 - 0.3\exp(-t)\,, \\ \beta(t) = u(t,x=1) = 0.2\,, \qquad t \geq 0\,. \end{cases} \qquad (4.3.3)$$

This problem refers to the perturbation of a uniform initial flow by a density at the end of the road which increases exponentially in time to the density $u = 0.5$.

The system of ordinary differential equations derived by the application of the collocation method is as follows:

$$\frac{du_i}{dt} = (2u_i - 1)\sum_{j=1}^{n} a_{ji}u_j + \eta u_i^2(1-u_i)\sum_{j=1}^{n} b_{ji}u_j$$

$$+ \eta u_i(2 - 3u_i) \left(\sum_{j=1}^{n} a_{ji} u_j \right)^2 , \tag{4.3.4}$$

for $i, j = 1, \ldots, n$, and where the coefficients $a_{ji} = a_{ji}^{(1)}$ and $b_{ji} := a_{ji}^{(2)}$ have been defined in Section 4.2.

The initial conditions are defined by Eq. (4.3.2):

$$u_i = 0.2 \,, \quad \text{for all} \quad i = 1, \ldots, n \,,$$

while the boundary conditions are implemented as indicated by Eq. (4.3.3).

We remark that rather than substituting u_1 with α and u_n with β, it is also possible to substitute the first and last equations by

$$\frac{du_1}{dt} = 0.3 \exp(-t) \,, \qquad \frac{du_n}{dt} = 0 \,. \tag{4.3.5}$$

The result of the simulation with $n = 11$ and $\eta = 0.2$ is shown in Figure 4.3.1.a which represents the density u versus space x aand time t. Simulations have been obtained using Lagrange-type polynomials. Specifically, Figure 4.3.1.b shows the density $u(t; x = 0.5)$ versus time t in the central node corresponding to an increasing number of nodes from $n = 7$ to $n = 15$. Finally, Figure 4.3.1.c shows $u(x, t)$ for various values of t varying from $t = 0$ to $t = 1$: $t = 0$ (solid line), $t = 0.25$ (dotted line), $t = 0.5$ (dotted-dashed line) $t = 0.75$ (double dotted-dashed line) and $t = 1$ (dashed line). Simulations have been performed using the program ***Traffic1***, see Appendix Section A.9.

Simulations show a smooth behavior without numerical instabilities, although the nodes number is small ($n = 11$).

Moreover, note how the Lagrange polynomials allow us to accurately reproduce the behavior at the domain boundaries. In the case of Figure 4.3.1.b numerical experiments show that even a small number of nodes is able to provide accurate results.

Analogous calculations can be developed for model (1.2.29), which uses the dimensionless flow q as a dependent variable. The model reported in Example 1.2.5 is written as follows:

$$\begin{cases} \dfrac{\partial u}{\partial t} = -\dfrac{\partial q}{\partial x} \,, \\[2ex] \dfrac{\partial q}{\partial t} = \left[(1 - 2u) + \eta u(2 - 3u) \dfrac{\partial q}{\partial x} \right] \dfrac{\partial q}{\partial x} + \eta u(1 - u) \dfrac{\partial^2 q}{\partial x^2} \,. \end{cases} \tag{4.3.6}$$

Let us consider the following problem.

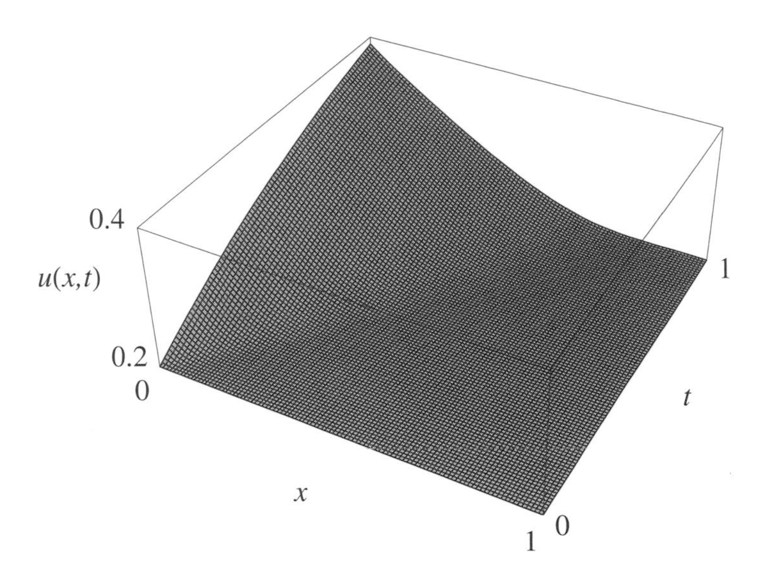

Figure 4.3.1.a - Density versus space and time.

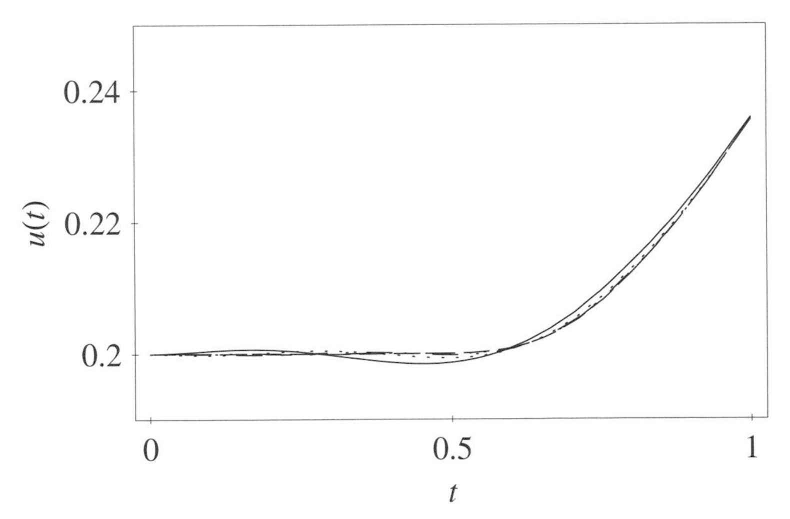

Figure 4.3.1.b - Density versus time in the central node.

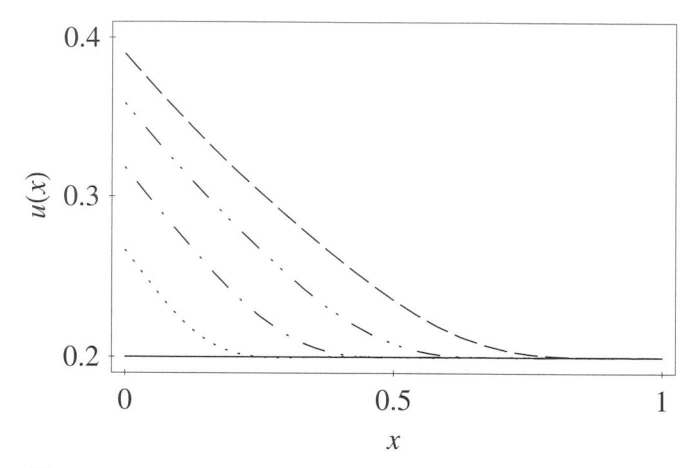

Figure 4.3.1.c - Density versus space at fixed times.

Problem 4.3.2. *The initial-boundary value problem is stated by linking Eq. (4.3.6) to the following initial and boundary conditions:*

$$\begin{cases} u_0(x) = u(t=0,x) = 0.2\,, \\ q_0(x) = q(t=0,x) = 0.16\,, \end{cases} \tag{4.3.7}$$

and

$$\begin{cases} \alpha(t) = q(t, x=0) = 0.16 = \text{Constant}\,, \\ \beta(t) = q(t, x=1) = 0.16 + 0.1\sin(5t)\,. \end{cases} \tag{4.3.8}$$

The above problem is interesting for the application considering that the measurement of the flux is relatively more accurate than the one of the density. Moreover, the statement of boundary conditions for the flux is useful when the road is dealt with as part of a network.

In detail, the application of the method generates the following system of ordinary differential equations:

$$\begin{cases} \dfrac{du_1}{dt} = -\sum_{j=1}^{n} a_{j1}^{(1)} q_j \,, \\ \quad \vdots \\ \dfrac{du_i}{dt} = -\sum_{j=1}^{n} a_{ji}^{(1)} q_j \,, \\ \quad \vdots \\ \dfrac{du_n}{dt} = -\sum_{j=1}^{n} a_{jn}^{(1)} q_j \,, \\ \dfrac{dq_1}{dt} = 0 \,, \\ \quad \vdots \\ \dfrac{dq_i}{dt} = \left[(1-2u_i) + \eta u_i (2-3u_i) \sum_{j=1}^{n} a_{ji}^{(1)} q_j \right] \\ \qquad \times \sum_{j=1}^{n} a_{ji}^{(1)} q_j + \eta u_i (1-u_i) \sum_{j=1}^{n} a_{ji}^{(2)} q_j \,, \\ \quad \vdots \\ \dfrac{dq_n}{dt} = 0.5 \cos(5t) \,, \end{cases} \tag{4.3.9}$$

with initial conditions

$$\begin{cases} u_1(0) = \cdots = u_i(0) = \cdots = u_n(0) = 0.2 \,, \\ q_1(0) = \cdots = q_i(0) = \cdots = q_n(0) = 0.16. \end{cases} \tag{4.3.10}$$

The result of the simulation with $n = 11$ and $\eta = 0.1$ is shown in Figures 4.3.2.a and 4.3.2.b which represent respectively the density u and the flow q versus space x and time t. Simulations have been obtained using Lagrange-type polynomials. Finally, in Figures 4.3.2.c and 4.3.2.d, $u(x,t)$ and $q(x,t)$ are respectively represented for various values of t varying from $t = 0$ to $t = 1$, namely $t = 0$ (solid line), $t = 0.25$ (dotted line), $t = 0.5$ (dotted-dashed line) $t = 0.75$ (double dotted-dashed line) and $t = 1$ (dashed line).

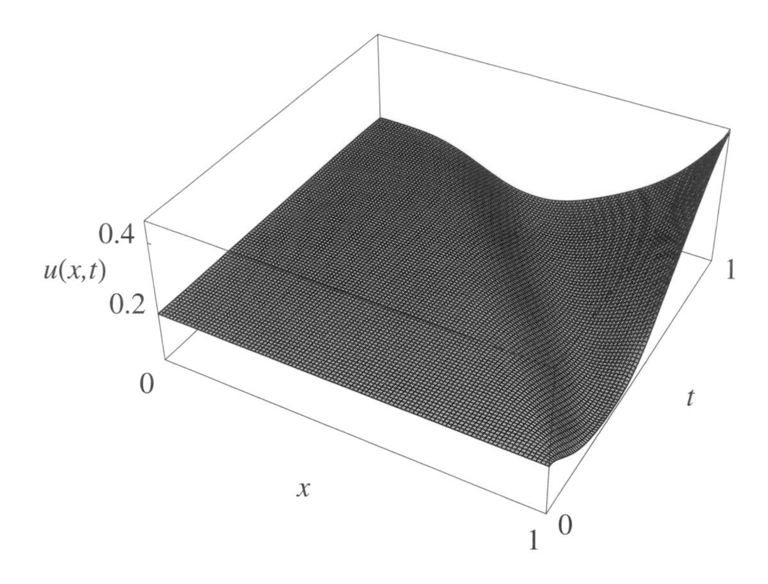

Figure 4.3.2.a - Density versus space and time.

Simulations are performed with the program ***Traffic2*** (Appendix, Section A.10). The figures confirm the same very accurate results observed previously. In particular, the spatial behaviors for different times shown in Figures 4.3.2.a and 4.3.2.b highlight the regularity of the simulations.

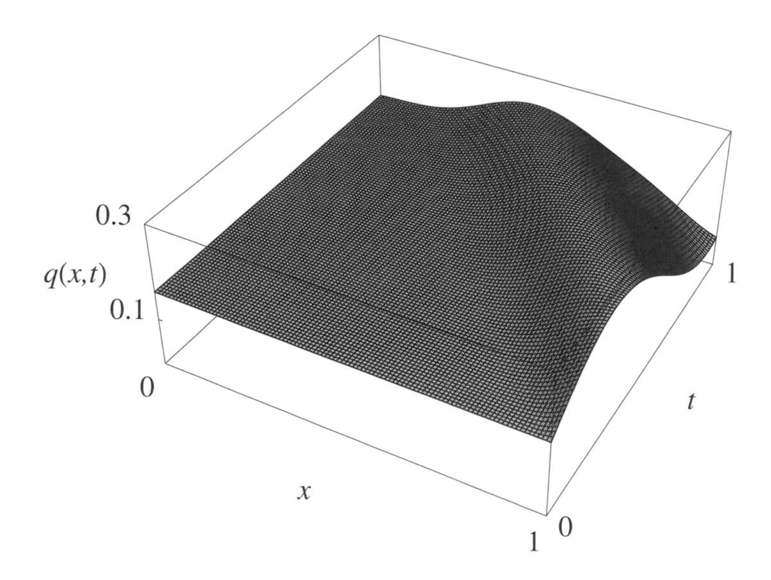

Figure 4.3.2.b - Flow versus space and time.

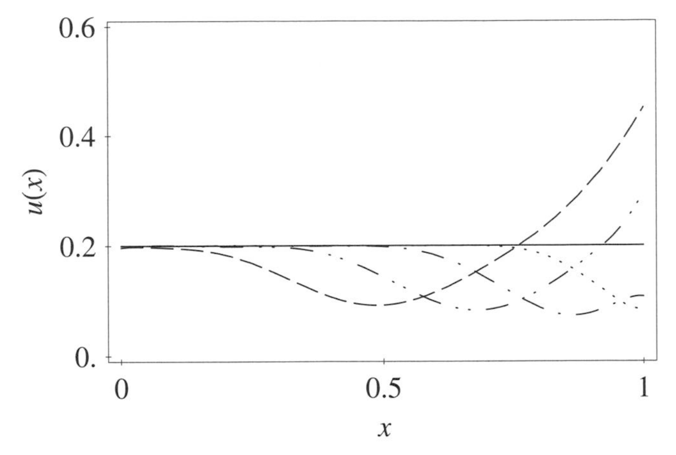

Figure 4.3.2.c - Density versus space at fixed times.

Note that the Neumann problem is also physically meaningful in the context of traffic flow models. For instance, the assumption

$$\frac{\partial u}{\partial x}(t,0) = \frac{\partial u}{\partial x}(t,1) = 0\,, \tag{4.3.11}$$

corresponds to the absence of gradients at the boundaries. This situation may be induced by tollgates. The above conditions can be implemented to Eq. (4.3.1), while a similar reasoning can be applied to model (4.3.6).

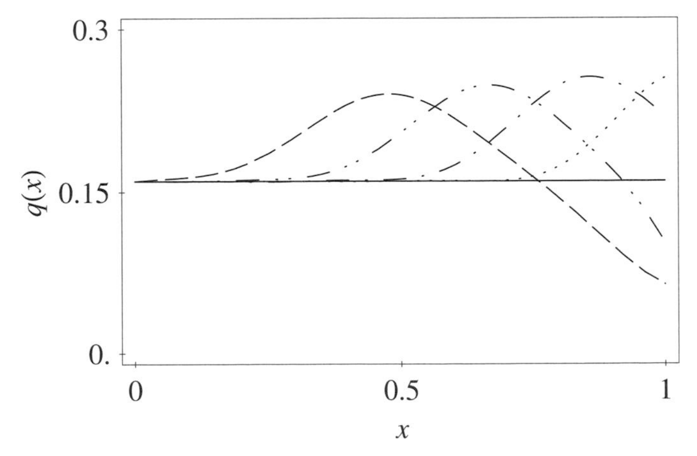

Figure 4.3.2.d - Flow versus space at fixed time.

4.4 Nonlinear Diffusion Models with Neumann and Robin Boundary Conditions

This section deals with the solution of the nonlinear heat diffusion model of Chapter 1 (Example 1.3.1). The model is rewritten here assuming that the diffusion coefficient $k = k(u)$ is given by $k = u(1-u)$. In this case the diffusion equation (1.3.5) is written

$$\frac{\partial u}{\partial t} = u(1-u)\frac{\partial^2 u}{\partial x^2} + (1-2u)\left(\frac{\partial u}{\partial x}\right)^2, \tag{4.4.1}$$

where u is the dimensionless concentration of some chemical streaming along a one-dimensional channel; the dependent variables are the time t and the space coordinate $x \in [0,1]$.

Referring to the above model, consider the following problem.

Problem (Neumann) 4.4.1. *The initial-boundary value problem is stated by linking Eq. (4.4.1) to the following initial conditions:*

$$u_0(x) = 4x^2(x-1)^2, \tag{4.4.2}$$

and Neumann boundary conditions:

$$\gamma(t) = \frac{\partial u}{\partial x}(t,0) = 0, \qquad \delta(t) = \frac{\partial u}{\partial x}(t,1) = 0. \tag{4.4.3}$$

The use of the method proposed in Section 4.2, i.e., the application of (4.2.12) and (4.2.13) to the above problem, leads to the following system:

$$\begin{cases} u_1 = \dfrac{1}{a_{n1}^{(1)}a_{1n}^{(1)} - a_{11}^{(1)}a_{nn}^{(1)}}\left[a_{nn}^{(1)}\displaystyle\sum_{j=2}^{n-1} a_{j1}^{(1)}u_j - a_{n1}^{(1)}\sum_{j=2}^{n-1} a_{jn}^{(1)}u_j\right], \\ \quad\vdots \\ \dfrac{du_i}{dt} = u_i(1-u_i)\displaystyle\sum_{j=1}^{n} a_{ji}^{(2)}u_j + (1-2u_i)\left(\sum_{j=1}^{n} a_{ji}^{(1)}u_j\right)^2, \\ \quad\vdots \\ u_n = \dfrac{1}{a_{11}^{(1)}a_{nn}^{(1)} - a_{n1}^{(1)}a_{1n}^{(1)}}\left[a_{1n}^{(1)}\displaystyle\sum_{j=2}^{n-1} a_{j1}^{(1)}u_j - a_{11}^{(1)}\sum_{j=2}^{n-1} a_{jn}^{(1)}u_j\right], \end{cases} \tag{4.4.4}$$

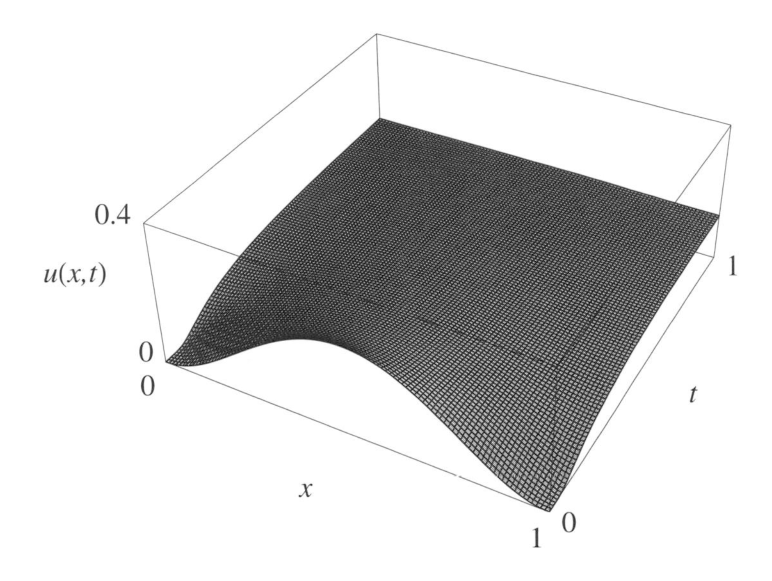

Figure 4.4.1.a - u versus space and time.

with the initial condition

$$\begin{cases} u_2(0) = 4\,x_2^2(x_2-1)^2\,, \\ \quad\vdots \\ u_i(0) = 4\,x_i^2(x_i-1)^2\,, \\ \quad\vdots \\ u_{n-1}(0) = 4\,x_{n-1}^2(x_{n-1}-1)^2\,. \end{cases} \tag{4.4.5}$$

The results of the simulations for $n = 21$ nodes using interpolation by Lagrange polynomials are shown in Figures 4.4.1.a and 4.4.1.b, showing, respectively, u versus time and space and u versus space for fixed values of t from $t = 0$ to $t = 1$: $t = 0$ (solid line), $t = 0.25$ (dotted line), $t = 0.5$ (dotted-dashed line) $t = 0.75$ (double dot-dashed line) and $t = 1$ (dashed line). Simulations are obtained with the help of the program ***DiffNeu*** (Appendix, Section A.11).

It is also possible to investigate the following problem involving Robin boundary conditions (with reference to Problem 4.2.2).

Problem (Robin) 4.4.2. *The initial-boundary value problem is stated by linking Eq. (4.4.1) to the initial conditions (4.4.2) and to the boundary*

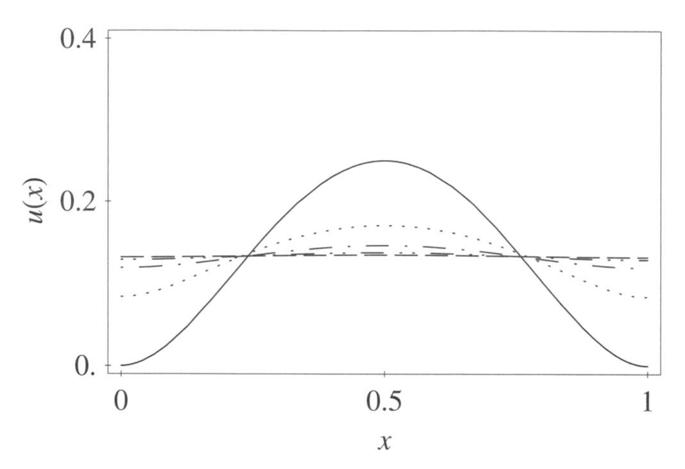

Figure 4.4.1.b - u versus space for fixed times.

conditions

$$\begin{cases} \alpha(t) + \delta(t) = 0\,, \\ \beta(t) + \gamma(t) = 0\,. \end{cases} \tag{4.4.6}$$

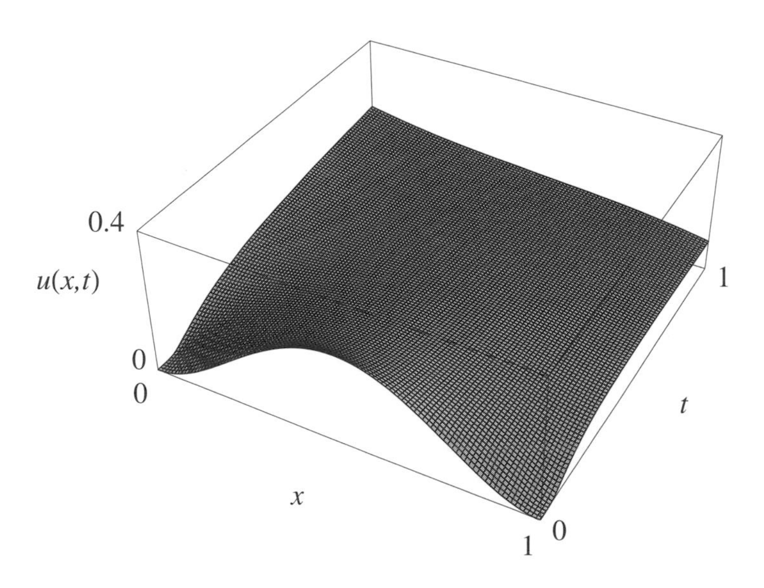

Figure 4.4.2.a - u versus space and time.

Performing simulations with $n = 21$ nodes, again using Lagrange interpolation we show, respectively, u versus time and space and u versus space in Figures 4.4.2.a and 4.4.2.b, where the last figure shows $u(x,t)$ for various value of t, from $t = 0$ to $t = 1$: $t = 0$ (solid line), $t = 0.25$ (dotted line), $t = 0.5$ (dotted-dashed line) $t = 0.75$ (double dot-dashed line) and $t = 1$ (dashed line). The simulations have been performed with the help of the program ***DiffRob*** (Appendix, Section A.12).

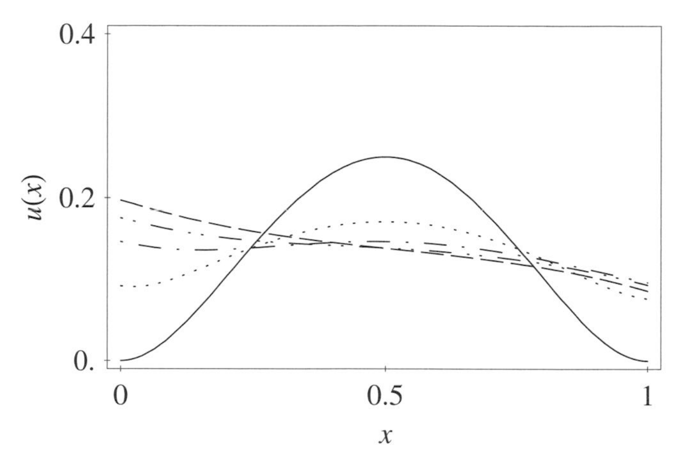

Figure 4.4.2.b - u versus space at fixed times.

4.5 Problems with Known Analytic Solutions

This section deals with the application of the method to problems where an analytic solution is available. Analytic solutions can contribute to identifying the number of nodes which are necessary to obtain an accurate solution. This matter is also discussed in the next section. Generally, analytic solutions are not available for nonlinear problems; however, the addition of a suitable source term to differential models can be technically organized to extract, as will be shown later, an analytic solution.

When an analytic solution is available, then it is possible to obtain a direct estimate of the errors generated by the application of the method; namely, the distance, defined by a suitable norm, between the computational and the analytic solution. The analysis can be developed only for the Dirichlet problem solved by application of both the sinc and Lagrange interpolations. This will allow at least in some cases a comparison between the use of the above two different techniques. The analysis is referred to the following.

Problem (Linear convection diffusion model with source) 4.5.1. *Let us consider the following model:*

$$\frac{\partial u}{\partial t} + \frac{\partial u}{\partial x} = \frac{\partial^2 u}{\partial x^2} + s(t,x)\,, \tag{4.5.1}$$

where $s(t,x)$ is a source term. The problem is stated with initial condition

$$u_0(x) = \exp[(1-x)^2 - 1]\,, \qquad \forall\, x \in [0,1]\,, \tag{4.5.2}$$

and boundary conditions

$$\alpha(t) = \exp(-t), \quad \beta(t) = \exp(-t-1) \qquad \forall\, t \in [0,1]\,. \tag{4.5.3}$$

In general, analytic solutions are not known referring to the above problem unless the source term is given by a very special expression. For instance, the following analytic solution:

$$u(t,x) = \exp[(1-x)^2 - t - 1]\,, \tag{4.5.4}$$

corresponds to the following source term:

$$s(t,x) = -\exp[x^2 - 2x - t][4x^2 - 10x + 9]\,. \tag{4.5.5}$$

The solution to Problem 4.5.1 with source term (4.5.5) can be obtained applying the method described in Section 4.2. The results which follow have been obtained using $n = 21$ nodes. The simulations are obtained with the program ***ConvLaSi*** reported in the Appendix, Section A.13.

Figure 4.5.1.a shows both the analytical solution $u(x,t)$ and simulations by Lagrange polynomials which are indistinguishable. Figure 4.5.1.b shows the interpolation of (4.5.4) by sinc functions. One notes that, for sinc interpolation, an error is already developed in the interpolation and approximation of the initial datum (4.5.2). The detailed representation of this error is visualized in Figure 4.5.1.c.

Figures 4.5.1.d and 4.5.1.e illustrate the error representation for both the collocation method based on sinc functions (Figure 4.5.1.d) and Lagrange polynomials (Figure 4.5.1.e). The error can be technically estimated by taking the L_∞-norm which shows that, in the case of sinc interpolation, one has

$$\|u(t,x) - u^n(t,x)\|_\infty \le .2\,, \tag{4.5.7}$$

while in the case of Lagrange interpolation, the error is consistently smaller:

$$\|u(t,x) - u^n(t,x)\|_\infty \le 5 * 10^{-8}\,. \tag{4.5.8}$$

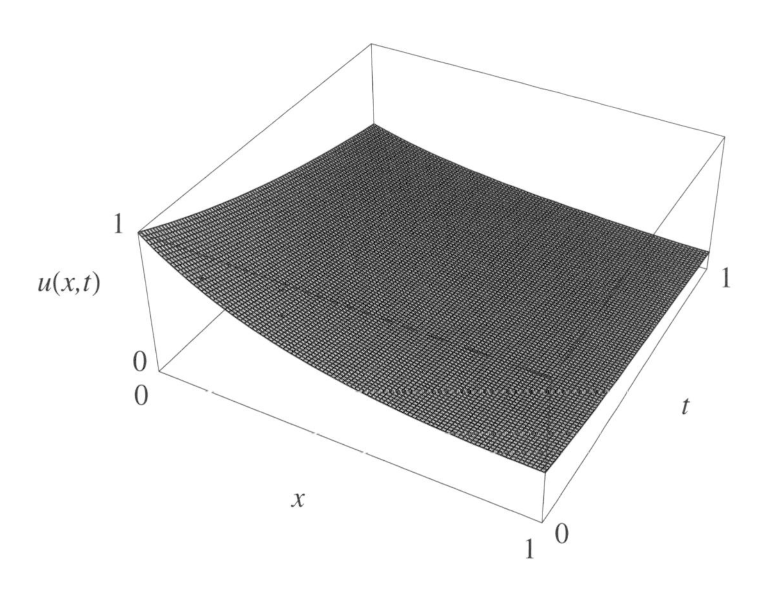

Figure 4.5.1.a - Simulation with Lagrange interpolation with 21 nodes.

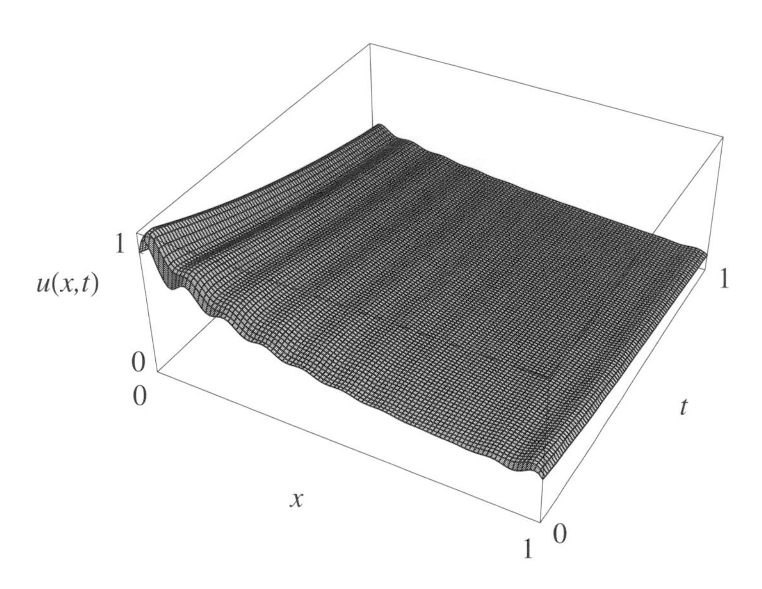

Figure 4.5.1.b - Simulation with sinc functions (21 nodes).

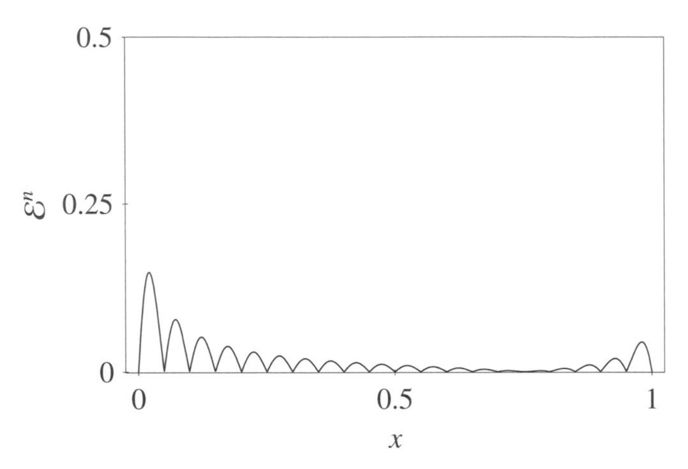

Figure 4.5.1.c - Error in the sinc representation of the initial datum.

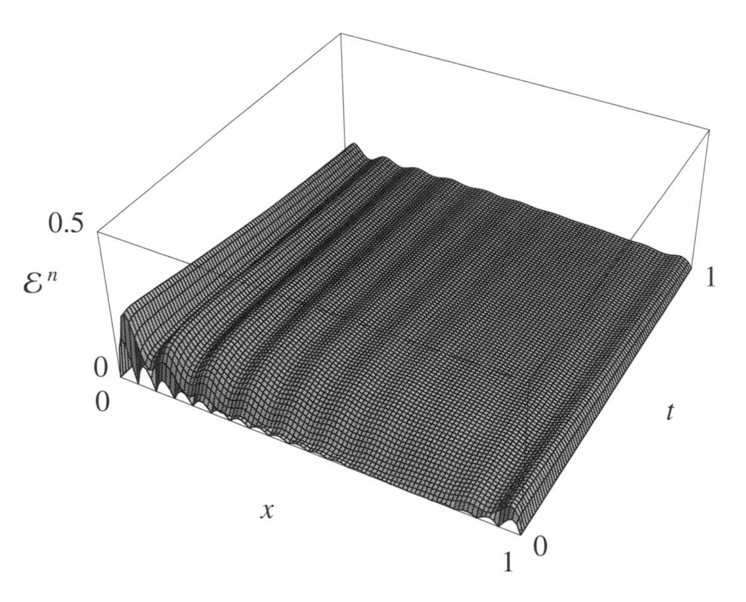

Figure 4.5.1.d - Error in the simulation by sinc functions.

These two values confirm how sinc interpolation is not suitable when boundary condition are different from zero. In this case, the rule is always to adopt a Lagrange interpolation with Chebychev-type collocation.

The above results can be improved, up to a certain extent, by increasing the number of collocation points. This matter will be discussed in the next

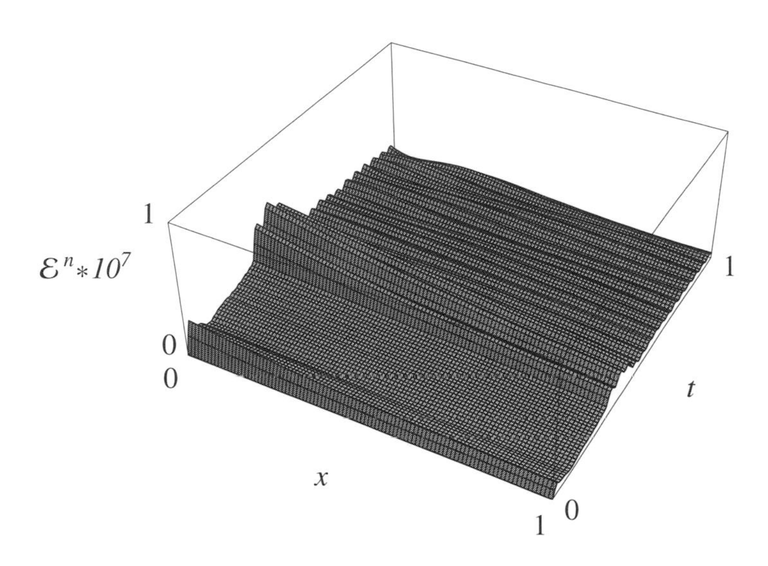

Figure 4.5.1.e - Error in the Lagrange interpolation of the initial datum.

section, bearing in mind that the above conclusions, which are related to this specific application, do not provide general rules. They have to be practically verified in each case. Indeed, the advantage of one interpolation with respect to the other depends also on the type of initial and boundary conditions. Analytic solutions can be useful towards this type of analysis.

4.6 On the Selection of the Number of Nodes

The various applications in this chapter have shown how generalized collocation methods can be applied to solve initial-boundary value problems for nonlinear partial differential equations with linear, Dirichlet, Neumann, and Robin boundary conditions. Comparisons with analytic solutions derived by adding an artificial source term have shown that accurate solutions can be obtained by a sufficiently large number of nodes.

Still, various technical problems need to be solved; specifically, the following:

i) Selection of the interpolating functions.

ii) Selection of the number of interpolating points.

Some heuristic reasoning is offered in what follows with the aim of providing some useful technical indications for dealing with the above problems at a practical level.

Referring to item i), the sinc interpolation has been shown to be suitable for wavy, in space, solutions. However, the straightforward use of the above interpolation is limited to the Dirichlet problem. Therefore the reasoning proposed in what follows refers to both types of interpolations for the Dirichlet problem and only to Lagrange interpolation for the other problems.

Relatively more delicate is the second point considering that increasing the number of nodes improves the interpolation of the dependent variable: for n tending to infinity, the error tends to zero. On the other hand, n also corresponds to the number of ordinary differential equations which have to be integrated over time. If h is the time step to be used to control the error induced by integration of the system of ordinary differential equations, one has $n \nearrow \Longrightarrow h \searrow ; \; n \to \infty \Longrightarrow h \to 0$.

Moreover, very small values of h may lead to excessive computational times and to uncontrolled computational errors induced by computation of the ratio of very small quantities. In other words $\mathcal{E}^n$ blows up for $n \to \infty$.

We stress that the use of ***Mathematica***® overcomes several technical problems and, specifically, the selection of h for a given system of equations. The algorithms of the solver of ordinary differential equations are able to relate h to the specific characteristics of the equation. Therefore, the technical, however relevant, problem to be solved is the following:

Suppose that an error $\mathcal{E}$ is fixed a priori for a certain initial-boundary value problem. Which is the smallest number of nodes needed to solve the problem with an error bounded by $\mathcal{E}^n$?

It is a difficult problem which cannot be solved in general for a large variety of cases. On the other hand, a technical procedure can be proposed with reference to specific problems.

Consider a given initial-boundary value problem and let us introduce the distance $d_n = \|u^n - u^{n-1}\|$, that generally decreases with increasing n: $n \nearrow \Longrightarrow d_n \searrow$. Therefore, the assessment of the number n can be achieved, for a certain initial-boundary value problem looking at d_n: specifically selecting n in a way that $d_n < \mathcal{E}^n$, $\forall t \in [0, T]$. Moreover, still at a heuristic level, it can be argued that for a given model if the above procedure is developed for initial conditions with sharp variation in space and boundary conditions with sharp variation in time, then the scientific program can be used for relatively less challenging initial and boundary conditions.

Finally, let us focus on how collocation methods differ technically from finite difference methods that need discretization both of time and space variables. Space derivatives are computed using the discretization points which are, in the grid, close to the node where the dependent variable is computed. The resulting computational scheme is algebraic rather than differential. It can be solved in the linear case by simple techniques with computational complexity generally lower than the one of collocation methods. On the other hand, nonlinear problems need additional technical treatments such as local linearization or solution of nonlinear algebraic systems. Technical methods to compute space derivatives heavily depend on the mathematical structure of the equation, say elliptic, parabolic, or hyperbolic. Several schemes are reported in Chapter 3 of the book by Bellomo and Preziosi (1996) with reference, as already mentioned, to the structure of the equation. If the time variable is left continuous and space derivatives are computed as outlined above, the approach is that of the ***method of lines***, which is similar to the one developed in this book and is besides a relatively less general way of computing space derivatives.

4.7 Problems

PROBLEM 4.1

Solve, by using the program ***DiffNeu***, the Dirichlet problem, with time-dependent boundary conditions, for the sine-Gordon model (1.2.14).

PROBLEM 4.2

Consider the initial-boundary value problem for the KdV solitary wave model (1.2.9) with Dirichlet boundary conditions at $x = 0$ and $x = 1$ and Neumann boundary condition at $x = 0$. Show how this additional condition can be implemented.

PROBLEM 4.3

Solve, by using the program ***DiffRob***, the Robin problem, with time-dependent boundary conditions, for the sine-Gordon model (1.2.14).

PROBLEM 4.4

Consider the initial-boundary value problem for the KdV solitary wave model with Neumann boundary conditions at $x = 0$ and $x = 1$ and Dirichlet boundary conditions at $x = 0$. Show how this additional condition can be implemented.

PROBLEM 4.5

Solve the initial-boundary value problem for the traffic flow model with Neumann boundary conditions.

PROBLEM 4.6
Solve the initial-boundary value problem for the traffic flow model with Robin boundary conditions.

PROBLEM 4.7
Find a suitable source term for the traffic flow model which gives an analytic solution to Problem 4.3.1 with Dirichlet boundary conditions.

PROBLEM 4.8
Solve the initial-boundary value problem related to Problem 4.7 comparing analytic and computational solutions.

PROBLEM 4.9
Solve Problem 4.5.1 and compute $d_n = \|u^n - u^{n-1}\|$ for various values of n. Compare this result with $\mathcal{E}^n$.

PROBLEM 4.10
Develop the analysis proposed in Problem 4.7 for Neumann boundary conditions.

5

Initial-Boundary Value Problems in Two Space Dimensions

5.1 Introduction

Chapter 4 has shown how generalized collocation methods can be applied to the solution of several nonlinear initial-boundary value problems in one space dimension. The advantage of this method with respect to other techniques of applied mathematics is its rapid application which leads to a technically simple control of nonlinearities.

On the other hand, various technical difficulties arise when the method is applied to problems in a number of space dimensions larger than one. This is due to the fact that when the number of independent variables increases, the related system of ordinary differential equations generates a stiff system, so that severe approximation problems in the time integration are generated.

This chapter deals with the solution of initial-boundary value problems in two space dimensions. Technically, the application of the method is a straightforward generalization of the method presented in Chapter 3; on the other hand the above-mentioned stiffness problems must be carefully taken into account. Bearing all this in mind, this chapter is developed along seven additional sections.

- Section 5.2 concerns *the statement of mathematical problems for nonlinear models in two space dimensions*. The classical Dirichlet, Neumann, and mixed problems for space domains with regular boundary are stated.

- Section 5.3 presents the *description of the application of the generalized collocation methods* to the solution of the problems of Section 5.2. As we shall see, the application is a simple generalization of the methods described in Chapter 4, although various technical aspects must be dealt with.

• Section 5.4 concerns the *initial-boundary problems over a strip* as the domain of the space variable. It is shown how a mixed use of the sinc functions and Lagrange polynomials can be used for this class of problems. Specifically, sinc functions are used for the interpolation on the variable defined over the whole real line, while Lagrange polynomials are used for the interpolation on the variable defined on bounded domains.

• Section 5.5 presents the *application of the method to the parabolic heat equation* in two space dimensions. The computational solution of the Dirichlet problem for a linear model is compared with the associated analytic solution.

• Section 5.6 develops a second *application to a reaction-diffusion model*, due to Schnakenberg (1979), described in Chapter 1, Example 1.2.8. The model is linear in the diffusion term but includes a nonlinear source term.

• Section 5.7 provides a *critical analysis* on the contents of this chapter mainly addressed to analyze the applicability of the method to geometrically complex space domains.

• Section 5.8 proposes a number of *problems* suggested to the interested reader for applying the mathematical tools proposed in this chapter.

5.2 Mathematical Problems

This section discusses the statement of the initial-boundary value problems for partial differential equations in two space dimensions. The statement is first developed for scalar equations with second-order partial derivatives, and then technical generalizations are described for vector equations and models with higher-order derivatives.

Let us consider the following general class of equations:

$$\frac{\partial u}{\partial t} = f\left(t, \mathbf{x}, u, \frac{\partial u}{\partial x}, \frac{\partial u}{\partial y}, \frac{\partial^2 u}{\partial x \partial y}, \frac{\partial^2 u}{\partial x^2}, \frac{\partial^2 u}{\partial y^2}\right), \tag{5.2.1}$$

where

$$u = u(t, \mathbf{x}) : \quad [0, T] \times D \to \mathbb{R}, \qquad \mathbf{x} = (\mathrm{x}, \mathrm{y}) \tag{5.2.2}$$

is the dependent variable, and where the domain $D \subseteq \mathbb{R}^2$ of the space variables has a regular boundary ∂D. The function f is assumed to be smooth in its argument. After a suitable dimensional analysis, one can assume that $u \in [0, 1]$ and that the domain D is bounded as follows:

$$D \subseteq [0, 1] \times [0, 1]. \tag{5.2.3}$$

For instance, if the original spatial domain is a rectangle, with sides a and b along the respective directions x and y, the above normalization can be obtained easily by introducing the new spatial variables

$$x' = \frac{x}{a}, \qquad y' = \frac{y}{b}. \tag{5.2.4}$$

In the same way it is possible to deal with circular and elliptical domains. For instance, if the original domain is a circle centered at the origin with radius R, the problem can be recast into the polar coordinates $(\rho, \theta) \in [0, R] \times [0, 2\pi]$ and then normalized by the introduction of the new independent variables

$$\rho' = \frac{\rho}{R} \qquad \theta' = \frac{\theta}{2\pi} \tag{5.2.5}$$

such that $\rho' \in [0, 1]$ and $\theta' \in [0, 1]$.

The statement of the mathematical problems is first proposed assuming boundary ∂D of D regular so that a unit vector $\mathbf{n}$, directed towards the interior of D, can be defined in any point of ∂D. Bearing all the above in mind, consider the following problems.

Problem 5.2.1. *Consider the initial-boundary value problems for Eq. (5.2.1) with initial condition*

$$u(0, \mathbf{x}) = \phi(\mathbf{x}), \quad \mathbf{x} = (x, y) \in D, \tag{5.2.6}$$

and boundary conditions defined as follows:

i) ***Dirichlet***

$$\forall t \in [0, T], \quad \mathbf{x} \in \partial D: \quad u(t; \mathbf{x} \in \partial D) = \alpha(t, \mathbf{x}), \tag{5.2.7}$$

ii) ***Neumann***

$$\forall t \in [0, T], \quad \mathbf{x} \in \partial D: \quad \frac{\partial u}{\partial \mathbf{n}}(t; \mathbf{x} \in \partial D) = \gamma(t, \mathbf{x}), \tag{5.2.8}$$

where (5.2.7) and (5.2.8) are given as smooth function consistent, for $t = 0$, *with the initial condition (5.2.6).*

If the domain D satisfies the above regularity properties and is convex with respect to both axes, it can be decomposed into two subdomains in such a way that

$$D = D_1 \cup D_2, \quad \partial D = \partial D_1 \cup \partial D_2, \quad D_1 \cap D_2 = \partial D_1 \cap \partial D_2 = \emptyset, \tag{5.2.9}$$

where ∂D_1 and ∂D_2 belong to the contour of ∂D.

The statement of mixed-type problems refers to this decomposition as follows.

Problem (Mixed) 5.2.2. *Consider the initial-boundary value problem for Eq. (5.2.1) with initial condition (5.2.6) and Dirichlet boundary condition*

$$\forall t \in [0,T], \quad \mathbf{x} \in \partial D_1 : \quad u(t; \mathbf{x} \in \partial D_1) = \alpha_1(t, \mathbf{x}), \tag{5.2.10}$$

and Neumann boundary condition

$$\forall t \in [0,T], \quad \mathbf{x} \in \partial D_2 : \quad \frac{\partial u}{\partial \mathbf{n}}(t; \mathbf{x} \in \partial D_2) = \gamma_2(t, \mathbf{x}), \tag{5.2.11}$$

given as smooth functions of time consistent, for $t = 0$, with initial condition (5.2.6). Consistency of α_1 and γ_2 at the contact points of the two boundaries, ∂D_1 and ∂D_2, has to be verified.

Remark 5.2.1. *An analogous statement of the boundary condition is as follows:*

$$\forall t \in [0,T], \quad \mathbf{x} \in \partial D_1 : \quad \frac{\partial u}{\partial \mathbf{n}}(t; \mathbf{x} \in \partial D_1) = \gamma_1(t, \mathbf{x}), \tag{5.2.12}$$

and

$$\forall t \in [0,T], \quad \mathbf{x} \in \partial D_2 : \quad u(t; \mathbf{x} \in \partial D_2) = \alpha_2(t, \mathbf{x}). \tag{5.2.13}$$

The above statement can be technically particularized to specific problems when D is a rectangular domain.

Example 5.2.1
Rectangular Domains

Let us consider here the case of a rectangular domain D in $\mathbb{R}^2$ that, after suitable normalization of the independent variables, can be assumed to be $D = [0,1] \times [0,1]$. This means that D has been transformed into the unit square of $\mathbb{R}^2$. The boundary ∂D consists then in four distinct parts, namely,

$$\partial D = \Gamma_1 \cup \Gamma_2 \cup \Gamma_3 \cup \Gamma_4 \,, \tag{5.2.14}$$

where

$$\Gamma_1 = \{(x,0)\,,\, x \in [0,1]\}\,, \qquad \Gamma_2 = \{(x,1)\,,\, x \in [0,1]\}\,,$$

$$\Gamma_3 = \{(0,y)\,,\, y \in [0,1]\}\,, \qquad \Gamma_4 = \{(1,y)\,,\, y \in [0,1]\}\,.$$

Therefore, the ***Dirichlet boundary conditions*** can be restated, for any $t \in [0, T]$, as follows:

$$\forall x \in [0,1]: \quad u(t; x, 0) = \alpha(t, x), \quad u(t; x, 1) = \beta(t, x), \tag{5.2.15}$$

and

$$\forall y \in [0,1]: \quad u(t; 0, y) = \gamma(t, y), \quad u(t; 1, y) = \delta(t, y), \tag{5.2.16}$$

where α, β, γ, and δ are given smooth functions of their arguments consistent with the initial condition (5.2.6), which satisfy the consistency conditions

$$\begin{aligned} \alpha(t, 0) = \gamma(t, 0), \quad \alpha(t, 1) = \delta(t, 0), \\ \beta(t, 0) = \gamma(t, 1), \quad \beta(t, 1) = \delta(t, 1). \end{aligned} \tag{5.2.17}$$

In the same way, ***Neumann boundary conditions*** are stated, for any $t \in [0, T]$, as follows:

$$\forall x \in [0,1]: \quad \frac{\partial u}{\partial y}(t; x, 0) = \psi(t, x), \quad \frac{\partial u}{\partial y}(t; x, 1) = \zeta(t, x), \tag{5.2.18}$$

and

$$\forall y \in [0,1]: \quad \frac{\partial u}{\partial x}(t; 0, y) = \eta(t, y), \quad \frac{\partial u}{\partial x}(t; 1, y) = \lambda(t, y), \tag{5.2.19}$$

where ψ, ζ, η, and λ are given smooth functions of their arguments consistent, for $t = 0$, with the initial condition (5.2.6), which satisfy the consistency conditions

$$\begin{aligned} \psi(t, 0) = \eta(t, 0), \quad \psi(t, 1) = \lambda(t, 0), \\ \zeta(t, 0) = \eta(t, 1), \quad \zeta(t, 1) = \lambda(t, 1). \end{aligned} \tag{5.2.20}$$

Finally, ***mixed-type boundary conditions*** are stated as follows:

$$\forall x \in [0,1]: \quad u(t; x, 0) = \alpha(t, x), \quad u(t; x, 1) = \beta(t, x), \tag{5.2.21}$$

and

$$\forall y \in [0,1]: \quad \frac{\partial u}{\partial x}(t; 0, y) = \eta(0, y), \quad \frac{\partial u}{\partial x}(t; 1, y) = \lambda(t, y), \tag{5.2.22}$$

or else,

$$\forall x \in [0,1]: \quad \frac{\partial u}{\partial y}(t;x,0) = \psi(t,x)\,, \quad \frac{\partial u}{\partial y}(t;x,1) = \zeta(t,x)\,, \tag{5.2.23}$$

and

$$\forall y \in [0,1]: \quad u(t;0,y) = \gamma(t,y)\,, \quad u(t;1,y) = \delta(t,y)\,, \tag{5.2.24}$$

for any $t \in [0,T]$, where all α, β, γ, δ, ψ, ζ, η, and λ are given smooth functions of their respective arguments, consistent, for $t = 0$, with the initial condition (5.2.6), which satisfy the corresponding consistency conditions.

□

The above reasoning holds when the space domain is convex with respect to both axes. On the other hand, when this condition is not satisfied, additional analysis is needed, based on the decomposition of the domain into several subdomains. This topic is dealt with in Section 5.7.

The above statement of the problem refers to equations with second-order derivatives on both space variables. It requires boundary conditions on the closed contour ∂D, whereas equations with first-order derivatives need boundary conditions on open contours. Higher-order equations require additional conditions.

We also point out that systems of equations need boundary conditions corresponding to each equation consistent with its structure. The statement may differ for each equation: for instance, Dirichlet conditions for the first equation and Neumann conditions for the second.

5.3 Solution Methods

The solution method, as already mentioned, is a technical development of the one described in Chapter 4. The difference is that the spatial domain is now two dimensional.

Similarly to initial-boundary value problems in one dimension, Lagrange polynomials with a Chebychev collocation will be adopted for the interpolation (see Chapter 2, Section 2.2), though the use of sinc functions is technically possible. The application of the method is developed in the following steps.

1. The first step is the discretization of the spatial domain into the collocation

$$I_{xy} = \{x_1, \ldots, x_n\,;\, y_1, \ldots, y_m\}\,,$$

where equispaced collocation points or Chebychev interpolation are used.

2. The dependent variable $u = u(t, x, y)$ is interpolated and approximated by the values $u_{ij} = u(t, x_i, y_j)$ according to

$$u(t, x, y) \simeq u^{nm}(t, x, y) = \sum_{i=1}^{n} \sum_{j=1}^{m} L_i(x) L_j(y) u_{ij}(t), \tag{5.3.1}$$

where the expressions of the Lagrange polynomials $L_i(x)$ and $L_j(y)$ have been given in Chapter 2, Section 2.2.

3. Interpolation (5.3.1) is then used to approximate the spatial partial derivatives in the nodal points of I_{xy}. We recall the formulae given in Chapter 2, namely

$$\frac{\partial^r u}{\partial x^r}(t; x_i, y_j) \simeq \sum_{h=1}^{n} a_{hi}^{(r)} u_{hj}(t)\,, \tag{5.3.2}$$

$$\frac{\partial^r u}{\partial y^r}(t; x_i, y_j) \simeq \sum_{k=1}^{m} b_{kj}^{(r)} u_{ik}(t)\,, \tag{5.3.3}$$

and

$$\frac{\partial^2 u}{\partial x \partial y}(t; x_i, y_j) \simeq \sum_{h=1}^{n} \sum_{k=1}^{m} a_{hi}^{(1)} b_{kj}^{(1)} u_{hk}(t)\,, \tag{5.3.4}$$

where the coefficients $a_{hi}^{(r)}$ and $b_{kj}^{(r)}$ have been given in Section 2.2.

4. The initial-boundary value problem is then transformed into an initial value problem for ordinary differential equations that describes the evolution of the values $u_{ij}(t)$ of u in the nodes. The boundary conditions have to be enforced in the nodal points on the boundaries x_{1j}, x_{nj}, x_{i1}, and x_{im} $(i = 1, ..., n;\ j = 1, ..., m)$ of the domain D.

5. Finally, the solution of the initial-boundary value problem is obtained by the numerical solution of the initial value problem for ordinary differential equations and, then, by the interpolation of the solution by the method adopted in Step 2.

If the above procedure is applied to ***Problem 5.2.1***, with conditions (5.2.16)–(5.2.17), then one obtains the system of ordinary differential equations

$$\frac{du_{ij}}{dt} = f\left(t, x_i, y_j, u_{ij}, \sum_{h=1}^{n} a_{hi}^{(r)} u_{hj}(t), \sum_{k=1}^{m} b_{kj}^{(r)} u_{ik}(t), \sum_{h=1}^{n} \sum_{k=1}^{m} a_{hi}^{(1)} b_{kj}^{(1)} u_{hk}(t)\right) \tag{5.3.5}$$

supplemented with the initial conditions

$$u_{i1} = \alpha(t, x_i)\,, \qquad u_{im} = \beta(t, x_i)\,,$$

and

$$u_{1j} = \gamma(t, y_j)\,, \qquad u_{nj} = \delta(t, y_j)\,,$$

for any $i = 1, \ldots, n$, $j = 1, \ldots, m$, and $r = 1, 2$.

Remark 5.3.1. *Similar to one-dimensional domains, in this two-dimensional case, the equations concerning the boundary conditions can also be written in differential form. For example,*

$$\frac{du_{i1}}{dt} = \frac{d\alpha}{dt}\,, \qquad \frac{du_{im}}{dt} = \frac{d\beta}{dt}\,. \tag{5.3.6}$$

In the following example, we illustrate this approach with a particular simple case.

Example 5.3.1

Nonlinear Transport Diffusion Model

Consider the following nonlinear transport diffusion model which can be seen as a generalization of the model equation (3.2.1) introduced in Chapter 3:

$$\begin{aligned}\frac{\partial u}{\partial t} &= \eta_1(t,x,y)\frac{\partial u}{\partial x} + \eta_2(t,x,y)\frac{\partial u}{\partial y} + \mu_1(t,x,y)\frac{\partial^2 u}{\partial x^2} \\ &\quad + \mu_2(t,x,y)\frac{\partial^2 u}{\partial y^2} + \mu_3(t,x,y)\frac{\partial^2 u}{\partial x \partial y} + f\left(t, x, y, \frac{\partial u}{\partial y}, \frac{\partial u}{\partial y}\right).\end{aligned} \tag{5.3.7}$$

Here the functions η_i $(i = 1, 2)$, μ_j $(j = 1, 2, 3)$, and f are given continuous functions of their respective arguments. The dimensionless dependent variable is

$$u = u(t, \mathbf{x}) : \quad [0,1] \times D \to [0,1], \qquad \mathbf{x} = (x, y) \in D\,,$$

where $D \subseteq [0,1] \times [0,1]$ is a suitable domain of $\mathbb{R}^2$. For this specific problem, the steps described above yield a system of $n \times m$ ordinary differential equations corresponding to collocations (x_i, y_i), which can be readily written as follows:

$$\frac{du_{ij}}{dt} = \eta_1(t, x_i, y_j) \sum_{h=1}^{n} a_{hi}^{(1)} u_{hj}(t) + \eta_2(t, x_i, y_j) \sum_{k=1}^{m} b_{kj}^{(1)} u_{ik}(t)$$

$$+ \mu_1(t,x_i,y_j)\sum_{h=1}^{n} a_{hi}^{(2)} u_{hj}(t) + \mu_2(t,x_i,y_j)\sum_{k=1}^{m} b_{kj}^{(2)} u_{ik}(t)$$

$$+ \mu_3(t,x_i,y_j)\sum_{h=1}^{n}\sum_{k=1}^{m} a_{hi}^{(1)} b_{kj}^{(1)} u_{hk}(t)$$

$$+ f\left(t,x_i,y_j,\sum_{h=1}^{n} a_{hi}^{(1)} u_{hj}(t), \sum_{k=1}^{m} b_{kj}^{(1)} u_{ik}(t) u_{hj}(t)\right), \qquad (5.3.8)$$

for $i = 1,\ldots,n$ and $j = 1,\ldots,m$. The previous system, linked to initial conditions

$$u_{0ij} = u_0(x_i, y_j)$$

and the boundary conditions

$$\frac{du_{i1}}{dt} = \frac{d\alpha}{dt}, \quad \frac{du_{im}}{dt} = \frac{d\beta}{dt}, \quad \frac{du_{1j}}{dt} = \frac{d\gamma}{dt}, \quad \frac{du_{nj}}{dt} = \frac{d\delta}{dt},$$

can be numerically simulated using the standard techniques for ordinary differential equations. The Appendix reports programs to deal with this system, which is nonlinear as is the original partial differential model.

□

The solution method for ***Problem 5.2.2*** can be developed by the same paths as described for Problem 5.2.1. The only difference concerns the fact that **Neumann boundary conditions** must be imposed. Following the same approach described in Chapter 4, Section 4.2, it is necessary to set the consistency with the boundary conditions that yield the following linear algebraic system:

$$\begin{cases} a_{11}^{(1)} u_{1j} + a_{n1}^{(1)} u_{nj} = \eta - \displaystyle\sum_{k=2}^{n-1} a_{k1}^{(1)} u_{kj} \\ a_{1n}^{(1)} u_{1j} + a_{nn}^{(1)} u_{nj} = \lambda - \displaystyle\sum_{k=2}^{n-1} a_{kn}^{(1)} u_{kj} \\ b_{11}^{(1)} u_{i1} + b_{m1}^{(1)} u_{im} = \psi - \displaystyle\sum_{k=2}^{m-1} b_{k1}^{(1)} u_{ik} \\ b_{1m}^{(1)} u_{im} + b_{mm}^{(1)} u_{im} = \zeta - \displaystyle\sum_{k=2}^{m-1} b_{km}^{(1)} u_{ik}, \end{cases} \qquad (5.3.9)$$

(with $i = 2, ..., n-1; j = 1, ..., m$).

This algebraic system can be solved with respect to u_{1j}, u_{nj}, u_{i1}, and u_{im}, leading to the following:

$$
\begin{aligned}
u_{1j} = &- \frac{a_{nn}}{a_{n1}a_{1n} - a_{11}a_{nn}} \left(\eta - \sum_{k=2}^{n-1} a_{k1} u_{kj} \right) \\
&+ \frac{a_{n1}}{a_{n1}a_{1n} - a_{11}a_{nn}} \left(\lambda - \sum_{k=2}^{n-1} a_{kn} u_{kj} \right),
\end{aligned}
\tag{5.3.10}
$$

$$
\begin{aligned}
u_{nj} = &- \frac{a_{1n}}{a_{n1}a_{1n} - a_{11}a_{nn}} \left(\eta - \sum_{k=2}^{n-1} a_{k1} u_{kj} \right) \\
&- \frac{a_{11}}{a_{n1}a_{1n} - a_{11}a_{nn}} \left(\lambda - \sum_{k=2}^{n-1} a_{kn} u_{kj} \right),
\end{aligned}
\tag{5.3.11}
$$

$$
\begin{aligned}
u_{i1} = &- \frac{b_{mm}}{b_{m1}b_{1m} - b_{11}b_{mm}} \left(\psi - \sum_{k=2}^{m-1} b_{k1} u_{ik} \right) \\
&+ \frac{b_{m1}}{b_{m1}b_{1m} - b_{11}b_{mm}} \left(\zeta - \sum_{k=2}^{m-1} b_{km} u_{ik} \right),
\end{aligned}
\tag{5.3.12}
$$

and

$$
\begin{aligned}
u_{im} = &- \frac{b_{1m}}{b_{m1}b_{1m} - b_{11}b_{mm}} \left(\psi - \sum_{k=2}^{m-1} b_{k1} u_{ik} \right) \\
&+ \frac{b_{11}}{b_{m1}b_{1m} - b_{11}b_{mm}} \left(\zeta - \sum_{k=2}^{m-1} b_{km} u_{ik} \right).
\end{aligned}
\tag{5.3.13}
$$

The above relations can be introduced into system (5.3.5), ensuring the self-consistency of the system itself.

The case of mixed boundary conditions can be dealt with using the previous results; namely, by combining conveniently the expressions corresponding to Dirichlet and Neumann boundary conditions.

In principle, both interpolations, sinc or Lagrange, can be used. Considering that two interpolations are applied, one for each independent variable, then one can use different interpolations for each of them.

The technical generalization is also immediate for systems of partial differential equations. In this case the dependent variable is a vector, say $\mathbf{u}$. The solution technique leads to a system of equations for each component of $\mathbf{u}$.

5.4 Initial-Boundary Value Problems on a Slab

The solution method we have seen in Section 5.3 can be straightforwardly generalized for partial differential equations with dependent spatial variable defined over an infinite slab along the x-axis and bounded over the y-variable. If the variables are scaled, one can assume that

$$u = u(t, x, y) \quad : \quad [0, T] \times [0, 1] \times [0, 1] \to [0, 1]. \tag{5.4.1}$$

As already indicated in Chapter 2, the natural solution method consists in interpolating over the x-variable by sinc functions corresponding to equally spaced nodes, and over the y-variable by Lagrange polynomials corresponding to Chebychev collocations. Consequently, one has

$$u(t, x, y) \cong u^{nm}(t, x, y) = \sum_{i=-n}^{n} \sum_{j=1}^{m} S_i(x, h) L_j(y) u_{ij}(t) , \tag{5.4.2}$$

where, as in Section 5.3, $u_{ij}(t) = u(t, x_i, y_j)$.

The discrete dynamical system is then obtained in the same way as in Section 5.3, the only difference being the coefficients $a_{hk}^{(r)}$ are obtained from the sinc functions for interpolation over the x-axis and by Lagrange polynomials for interpolation over the y-axis.

The application of the method can be shown with the solution of the nonlinear heat diffusion model we have seen in Chapter 1. Specifically, consider the following model in two space dimensions:

$$\frac{\partial u}{\partial t} = \frac{\partial}{\partial x}\left[u(1-u)\frac{\partial u}{\partial x}\right] + \frac{\partial}{\partial y}\left[u(1-u)\frac{\partial u}{\partial y}\right], \tag{5.4.3}$$

where $u = u(t, x, y)$ represents the dimensionless dependent variable.

The above model can be explicitly written as follows:

$$\frac{\partial u}{\partial t} = (1-2u)\left[\left(\frac{\partial u}{\partial x}\right)^2 + \left(\frac{\partial u}{\partial y}\right)^2\right] + u(1-u)\left[\frac{\partial^2 u}{\partial x^2} + \frac{\partial^2 u}{\partial y^2}\right]. \tag{5.4.4}$$

The model can be implemented, in the application dealt with in this section below, with the initial condition

$$u_0(x, y) = u(t = 0, x, y) = e^{-ax^2} \tag{5.4.5}$$

that does not depend on the y-variable and with boundary conditions

$$\lim_{|x|\to\infty} u(t,x,y) = 0, \quad u(t,x,0) = u(t,x,1) = e^{-ax^2} e^{-bt}. \tag{5.4.6}$$

Equation (5.4.4) can be discretized as follows:

$$\begin{aligned}\frac{du_{ij}}{dt} =& (1-2u_{ij})\left[\left(\sum_{h=1}^{n} a_{hi}^{(1)} u_{hj}(t)\right)^2 + \left(\sum_{k=1}^{m} b_{kj}^{(1)} u_{ik}(t)\right)^2\right] \\ &+ u_{ij}(1-u_{ij})\left[\sum_{h=1}^{n} a_{hi}^{(2)} u_{hj}(t) + \sum_{k=1}^{m} b_{kj}^{(2)} u_{ik}(t)\right]\end{aligned} \tag{5.4.7}$$

with $i = 2, ..., n-1; \quad j = 2, ..., m-1.$

The initial condition (5.4.5) corresponds to

$$u_{ij}(t=0) = e^{-ax_i^2} \quad (i = 1, ..., n), \tag{5.4.8}$$

while the boundary conditions (5.4.6) are described according to

$$u_{1j} = u_{nj} = 0 \qquad (j = 1, ..., m), \tag{5.4.9}$$

$$u_{i2} = u_{im-1} = e^{-ax_i^2} e^{-bt} \qquad (i = 1, ..., n). \tag{5.4.10}$$

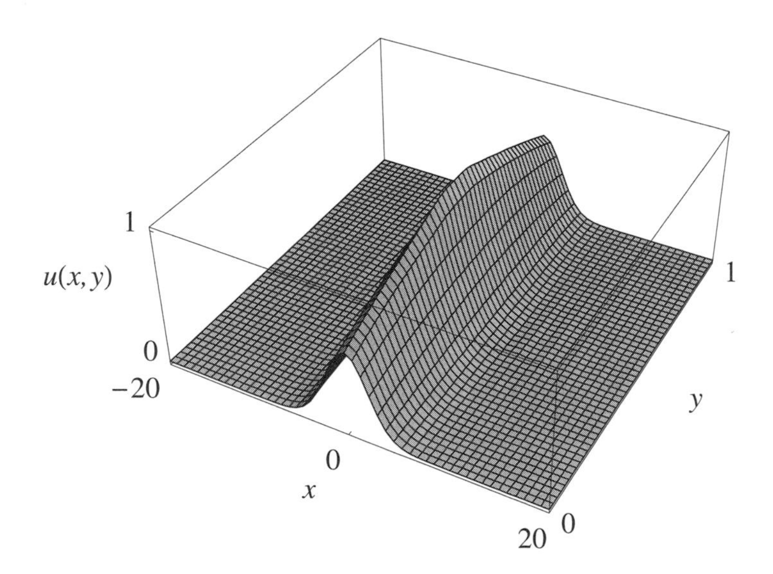

Figure 5.4.1 - u versus time and space: $a = 0.1$, $b = 1$, $t = 0.5$.

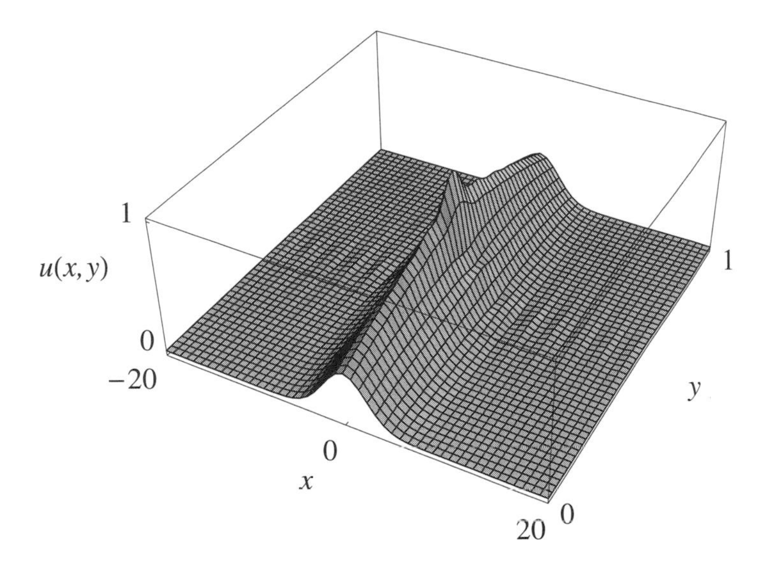

Figure 5.4.2 - u versus time and space: $a = 0.1$, $b = 1$, $t = 1$.

Figures 5.4.1 and 5.4.2 refer to the case with $a = 0.1$ and $b = 1$, and show the spatial behavior of the variable u at $t = 0.5$ and $t = 1$, respectively.

The program ***SlabTwoDim*** (see Appendix, Section A.14), has been used with 31×31 nodes. The method provides very accurate results characterized by surfaces that are regular and smooth. Very small spurious oscillations are detectable, in the central parts of the flat zones, that vanish when the number of nodes is increased.

5.5 Application: The Heat Equation over a Square

Let us consider the two-dimensional heat equation

$$\frac{\partial u}{\partial t} = \frac{1}{\pi^2}\left(\frac{\partial^2 u}{\partial x^2} + \frac{\partial^2 u}{\partial y^2}\right), \tag{5.5.1}$$

over the square domain $D = [0, 1] \times [0, 1]$, where the dimensionless dependent variable is

$$u = u(t, \mathbf{x}) : \quad [0, T] \times D \to [0, 1], \qquad \mathbf{x} = (x, y) \in D\,.$$

The partial differential equation (5.5.1) is linked to the following initial condition:

$$u(t=0,x,y) = u_0(x,y) = \cos\frac{\pi}{2}(x+y) + \sin\pi(x-y) \tag{5.5.2}$$

and boundary conditions

$$u(t,x,0) = e^{-\frac{t}{2}}\cos\frac{\pi}{2}x + e^{-2t}\sin\pi x\,, \tag{5.5.3}$$

$$u(t,x,1) = e^{-\frac{t}{2}}\cos\frac{\pi}{2}(x+1) + e^{-2t}\sin\pi(x-1)\,, \tag{5.5.4}$$

$$u(t,0,y) = e^{-\frac{t}{2}}\cos\frac{\pi}{2}y + e^{-2t}\sin\pi y\,, \tag{5.5.5}$$

and

$$u(t,1,y) = e^{-\frac{t}{2}}\cos\frac{\pi}{2}(1+y)e^{-2t}\sin\pi(1-y)\,. \tag{5.5.6}$$

The initial boundary problem (5.5.1)–(5.5.6) admits the following analytical solution:

$$u(t,x,y) = e^{-\frac{t}{2}}\cos\frac{\pi}{2}(x+y) + e^{-2t}\sin\pi(x-y). \tag{5.5.7}$$

The knowledge of the solution (5.5.7) allows a comparison with that obtained by the computational method and thus allows us to evaluate its accuracy.

Applying the procedure developed in Section 5.3, the corresponding system of ordinary differential equations is obtained:

$$\frac{du_{ij}}{dt} = \frac{1}{\pi^2}\left(\sum_{h=1}^{n} a_{hi}^{(2)} u_{hj}(t) + \sum_{k=1}^{m} b_{kj}^{(2)} u_{ik}(t)\right), \tag{5.5.8}$$

with $i = 1, ..., n$ and $j = 1, ..., m$; while the initial conditions, given by Eq. (5.5.2), are

$$u_{ij} = \cos\frac{\pi}{2}(x_i + y_j) + \sin\pi(x_i - y_j) \tag{5.5.9}$$

for $i = 1, \ldots, n$ and $j = 1, \ldots, m$. Finally, boundary conditions are implemented according to Eqs. (5.5.3–5.5.6).

Simulations have been obtained using Lagrange-type polynomials with $n = m = 11$ nodes. Figures 5.5.1 and 5.5.2 show the numerical solution $u(t, x, y)$ for $t = 0.5$ and $t = 1$, respectively. The smoothness of the solution, despite the small number of nodes, and the accuracy of the solution method

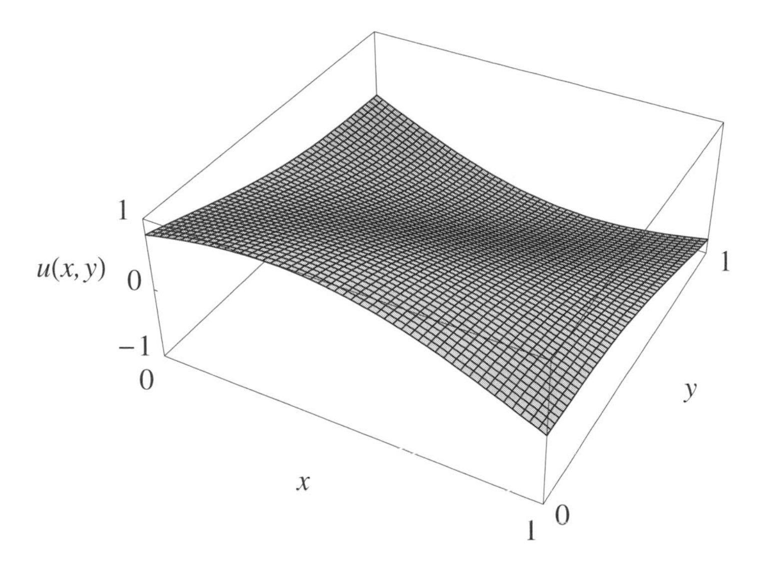

Figure 5.5.1 - u versus time and space: $t = 0.5$.

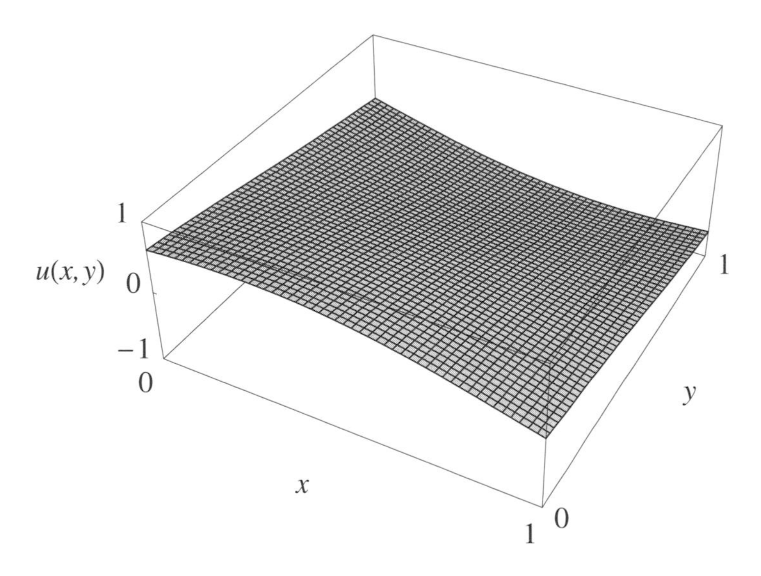

Figure 5.5.2 - u versus time and space: $t = 1$.

are visualized in Figure 5.5.3, which shows the error, at $t = 1$, with respect to the analytical solution (5.5.5).

It can be seen that the error is lower than 10^{-7}. Simulations have been performed using the program ***HeatTwoDim*** (see the Appendix, Section A.15).

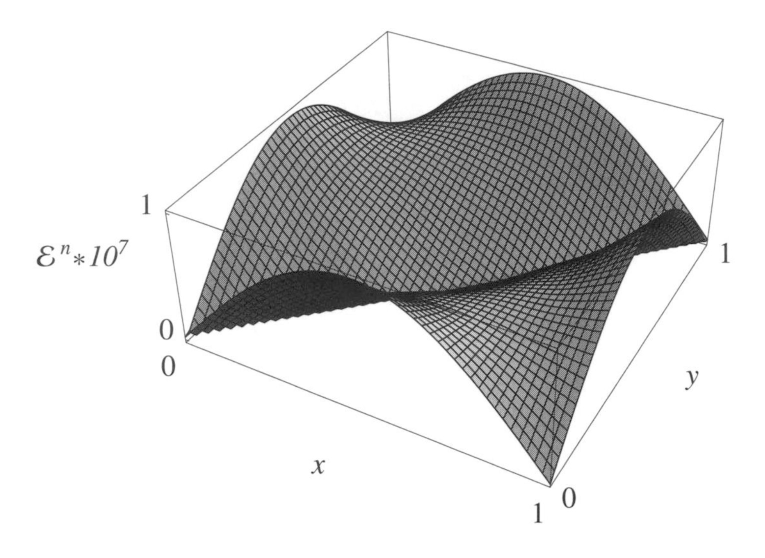

Figure 5.5.3 - Computational error versus time and space: $t = 1$.

5.6 Reaction-Diffusion Equations

Consider the Schnakenberg model of diffusion of chemicals described in Chapter 1, Example 1.2.8, which is written here again:

$$\begin{cases} \dfrac{\partial u}{\partial t} = \kappa(a - u + u^2 v) + \varepsilon\left(\dfrac{\partial^2 u}{\partial x^2} + \dfrac{\partial^2 u}{\partial y^2}\right), \\ \dfrac{\partial v}{\partial t} = \kappa(b - u^2 v) + \left(\dfrac{\partial^2 v}{\partial x^2} + \dfrac{\partial^2 v}{\partial y^2}\right), \end{cases} \tag{5.6.1}$$

where u and v are the two dependent variables

$$u = u(t, \mathbf{x}) : \quad [0,1] \times D \to [0,1], \qquad \mathbf{x} = (x, y) \in D,$$

and

$$v = v(t, \mathbf{x}) : \quad [0,1] \times D \to [0,1], \qquad \mathbf{x} = (x, y) \in D,$$

where $D = [0,1] \times [0,1]$.

The following parameters: $\kappa = 100$, $a = 0.1305$, $b = 0.7695$, and $\varepsilon = 0.05$, are used for the simulations (see Hundsdorfer and Verwer (2003)) for an initial boundary problem stated by linking the above model to the two dependent variables with the following initial conditions:

$$u(0,x,y) = a + b + \frac{1}{1000} e^{-100[(x-\frac{1}{3})^2 + (y-\frac{1}{2})^2]}\,, \tag{5.6.2}$$

and

$$v(0,x,y) = \frac{b}{(a+b)^2}\,, \tag{5.6.3}$$

while boundary conditions are assumed to be given by

$$\begin{cases} \dfrac{\partial u}{\partial x}(t,0,y) = \dfrac{\partial u}{\partial x}(t,1,y) = 0\,, \\ \dfrac{\partial v}{\partial x}(t,0,y) = \dfrac{\partial v}{\partial x}(t,1,y) = 0\,, \end{cases} \tag{5.6.4}$$

and

$$\begin{cases} \dfrac{\partial u}{\partial y}(t,x,0) = \dfrac{\partial u}{\partial y}(t,x,1) = 0\,, \\ \dfrac{\partial v}{\partial y}(t,x,0) = \dfrac{\partial v}{\partial y}(t,x,1) = 0\,, \end{cases} \tag{5.6.5}$$

which correspond to homogeneous Neumann conditions.

The application of the method generates the following system of ordinary differential equations:

$$\begin{cases} \dfrac{du_{ij}}{dt} = \kappa(a - u_{ij} + u_{ij}^2 v_{ij}) + \varepsilon\left(\displaystyle\sum_{k=1}^{n} a_{ki}^{(2)} u_{kj} + \sum_{k=1}^{m} b_{kj}^{(2)} u_{ik}\right), \\ \dfrac{dv_{ij}}{dt} = \kappa(b - u_{ij}^2 v_{ij}) + \left(\displaystyle\sum_{k=1}^{n} a_{ki}^{(2)} v_{kj} + \sum_{k=1}^{m} b_{kj}^{(2)} v_{ik}\right), \end{cases} \tag{5.6.6}$$

where $i = 2, \ldots, n-1$ and $j = 2, \ldots, m-1$, and with the initial conditions given by (5.6.2) and (5.6.3).

The solution of the initial-boundary value problem requires that we express the values of the two dependent variables u and v in the boundary nodes. Technical calculations yield

$$
\begin{cases}
u_{1,j} = \dfrac{a_{nn}^{(1)} \sum_{k=2}^{n-1} a_{k1}^{(1)} u_{kj} - a_{n1}^{(1)} \sum_{k=2}^{n-1} a_{kn}^{(1)} u_{kj}}{a_{n1}^{(1)} a_{1n}^{(1)} - a_{11}^{(1)} a_{nn}^{(1)}}, \\[2ex]
u_{n,j} = -\dfrac{a_{1n}^{(1)} \sum_{k=2}^{n-1} a_{k1}^{(1)} u_{kj} - a_{11}^{(1)} \sum_{k=2}^{n-1} a_{kn}^{(1)} u_{kj}}{a_{n1}^{(1)} a_{1n}^{(1)} - a_{11}^{(1)} a_{nn}^{(1)}}, \\[2ex]
v_{1,j} = \dfrac{a_{nn}^{(1)} \sum_{k=2}^{n-1} a_{k1}^{(1)} v_{kj} - a_{n1}^{(1)} \sum_{k=2}^{n-1} a_{kn}^{(1)} v_{kj}}{a_{n1}^{(1)} a_{1n}^{(1)} - a_{11}^{(1)} a_{nn}^{(1)}}, \\[2ex]
v_{n,j} = -\dfrac{a_{1n}^{(1)} \sum_{k=2}^{n-1} a_{k1}^{(1)} v_{kj} - a_{11}^{(1)} \sum_{k=2}^{n-1} a_{kn}^{(1)} v_{kj}}{a_{n1}^{(1)} a_{1n}^{(1)} - a_{11}^{(1)} a_{nn}^{(1)}},
\end{cases}
\tag{5.6.7}
$$

for $j = 1, \ldots, m$, and

$$
\begin{cases}
u_{1,j} = \dfrac{b_{mm}^{(1)} \sum_{k=2}^{m-1} b_{k1}^{(1)} u_{kj} - b_{m1}^{(1)} \sum_{k=2}^{m-1} b_{km}^{(1)} u_{kj}}{b_{m1}^{(1)} b_{1m}^{(1)} - b_{11}^{(1)} b_{mm}^{(1)}}, \\[2ex]
u_{m,j} = -\dfrac{b_{1m}^{(1)} \sum_{k=2}^{m-1} b_{k1}^{(1)} u_{kj} - b_{11}^{(1)} \sum_{k=2}^{m-1} b_{km}^{(1)} u_{kj}}{b_{m1}^{(1)} b_{1m}^{(1)} - b_{11}^{(1)} b_{mm}^{(1)}}, \\[2ex]
v_{1,j} = \dfrac{b_{mm}^{(1)} \sum_{k=2}^{m-1} b_{k1}^{(1)} v_{kj} - b_{m1}^{(1)} \sum_{k=2}^{m-1} b_{km}^{(1)} v_{kj}}{b_{m1}^{(1)} b_{1m}^{(1)} - b_{11}^{(1)} b_{mm}^{(1)}}, \\[2ex]
v_{m,j} = -\dfrac{b_{1m}^{(1)} \sum_{k=2}^{m-1} b_{k1}^{(1)} v_{kj} - b_{11}^{(1)} \sum_{k=2}^{m-1} b_{km}^{(1)} v_{kj}}{b_{m1}^{(1)} b_{1m}^{(1)} - b_{11}^{(1)} b_{mm}^{(1)}},
\end{cases}
\tag{5.6.8}
$$

for $i = 2, \ldots, n-1$.

Simulations have been obtained using Lagrange polynomials with 51 nodes by the program ***EvolPatt***, Appendix, Section A.16.

Figures 5.6.1.a–5.6.3.a show the evolution of the dependent variable u at $t = 0.25$, $t = 0.5$, and $t = 1$, respectively. The typical feature of patterns generated by the Schnakenberg model, that is, the amplification and the migration of the bumps, is plainly recovered by the simulations. This characteristic dynamics is even more evident in the representation by contour lines shown in the corresponding Figures 5.6.1.b–5.6.3.b, where such lines join points of domain having the same value of population, u.

Note that the above numerical simulations agree very well with the results reported by Hundsdorfer and Verwer (2003), where different and more complex numerical methods are used.

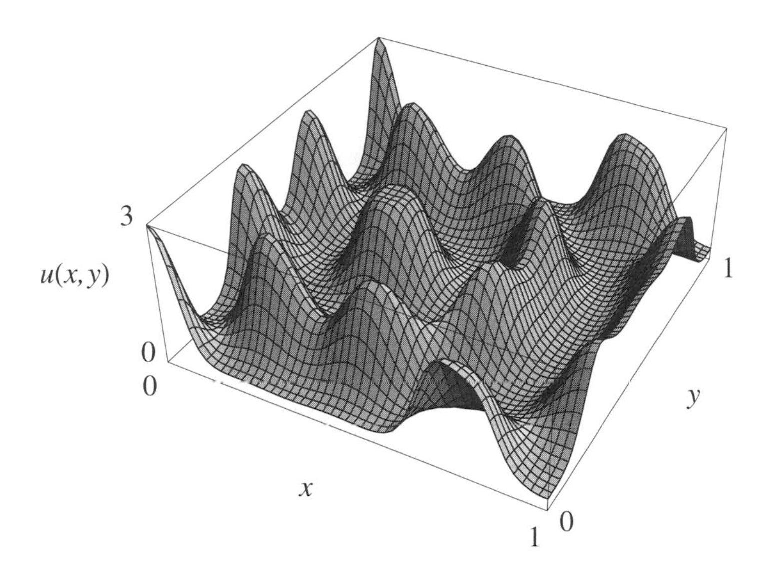

Figure 5.6.1.a - u versus space at $t = 0.25$.

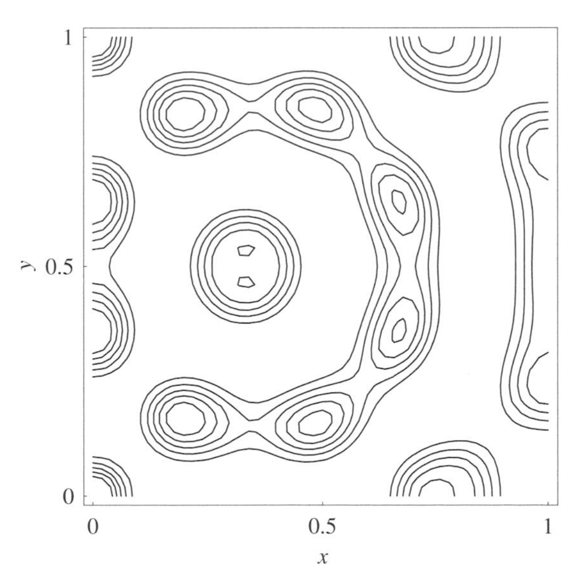

Figure 5.6.1.b - Contour lines corresponding to Figure 5.6.1.a.

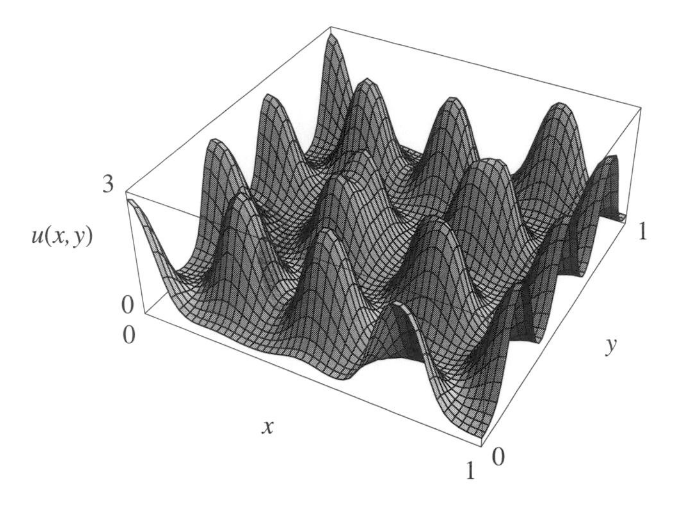

Figure 5.6.2.a - u versus space at $t = 0.5$.

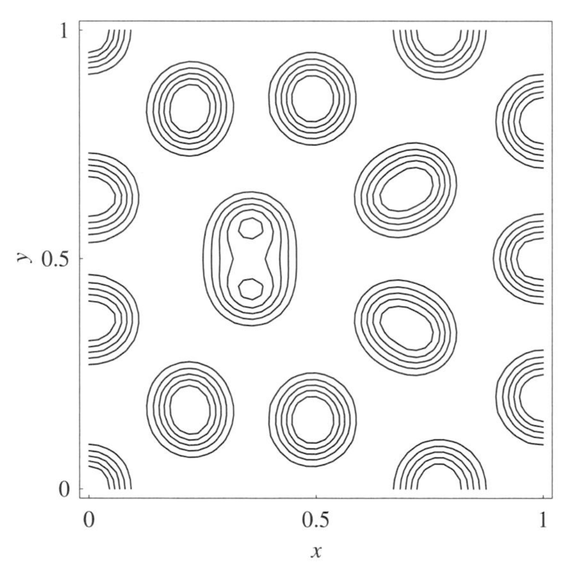

Figure 5.6.2.b - Contour lines corresponding to Figure 5.6.2.a.

5.7 Critical Analysis

Initial-boundary value problems in two space dimensions have been dealt with in this chapter. Various applications have shown how the method is flexible enough to deal efficiently with time-varying boundary conditions.

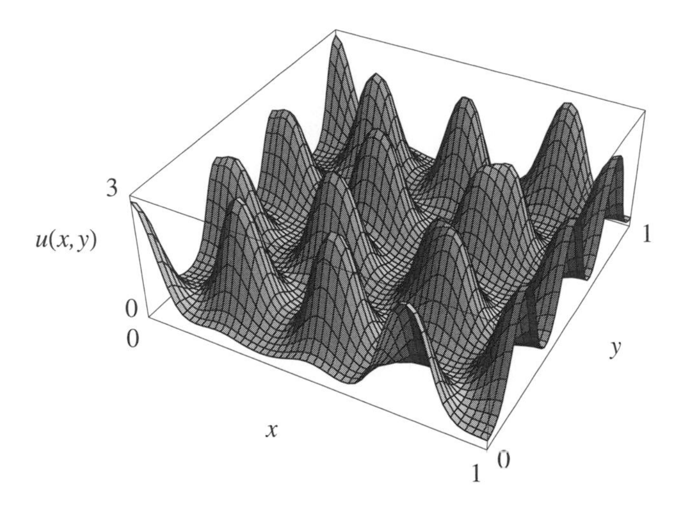

Figure 5.6.3.a - u versus space at $t = 1$.

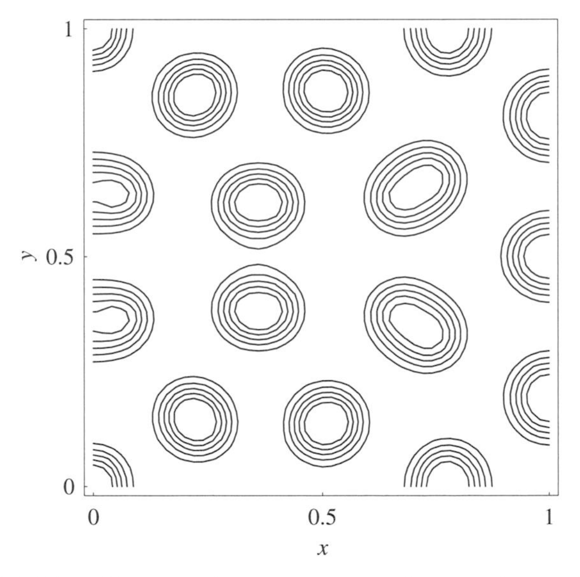

Figure 5.6.3.b - Contour lines corresponding to Figure 5.6.3.a.

The method provides accurate solutions, although the computational complexity is increased by dealing with problems in two space variables. On the other hand, a critical aspect is that the method can be straightforwardly applied in the case of simple geometries, say squares, circles, or rectangles, while additional work is needed for the more complex geometry of the domain D of the space variable.

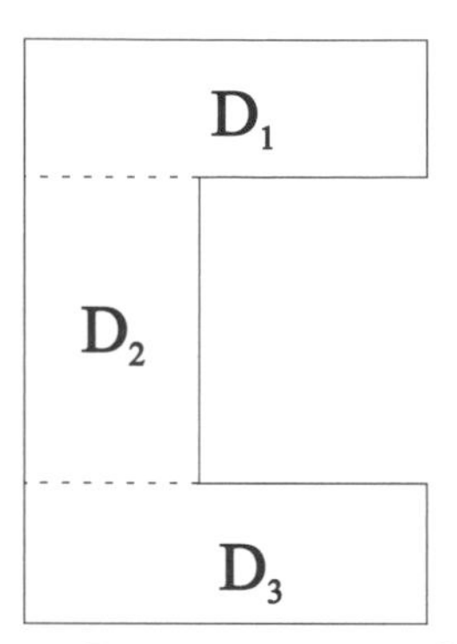

Figure 5.7.1 - Decomposition of domains.

This section analyzes the technical development of the method by taking into account different shapes of D.

The first problem is the case of domains D which can be decomposed as shown in the example of Figure 5.7.1 into different subdomains D_k still with a rectangular shape: $D = \bigcup_k D_k$. Subsequently, a method is developed as follows.

1. The space variables are discretized separately in each subdomain D_k. Some of the discretization points (specifically on their boundaries) are in common for two contiguous subdomains.
2. The mathematical model is then discretized into a system of $n = \sum_k p_k$ ordinary differential equations, where p_k is the number of nodes in each domain D_k.
3. Boundary conditions are imposed on the boundary ∂D of the space domain, while consistency conditions are imposed on the collocation points belonging to the common boundaries of each subdomain D_k.

The above procedure leads to a reduced system of n ordinary differential equations minus the number of equations corresponding to the collocation points on the boundaries of each domain D_k.

Another useful case refers to problems where the domain D is triangular, one of the sides is normalized over the interval $[0, 1]$, and scaling can also identify the position of the vertex.

Then a Chebychev collocation identifies the collocation at the side, while the collocation points are obtained by the intersections of the lines corresponding to the collocation on that sides obtained by joining the vertex opposite to that side.

The solution method then follows the same procedure we have seen before: discretizing the model into a system of ordinary differential equations, imposing boundary conditions on the collocation points of ∂D, and then solving the remaining system of ordinary differential equations. The above scheme can be used when the domain can be decomposed into several triangles D_k, say $D = \bigcup_{i=1}^{N} D_k$. Then, the solutions in each triangle can be linked, as we have seen in the case of decomposition into rectangular domains.

We note from these schemes that the generalization of the method does not always appear practical compared with other methods of applied mathematics. This remark appears to be even more consistent when the space variable is defined over a three-dimensional volume. In this case the formal generalization of the method can be technically developed by a three-dimensional interpolation. On the other hand, the computational complexity related to the large number of ordinary differential equations, of the order of $n \times m \times p$, cannot be dealt with properly.

5.8 Problems

PROBLEM 5.1

Develop a scientific program to deal with the nonlinear heat equation (5.4.3) on a slab. Compare the results with Figures 5.4.1 and 5.4.2.

PROBLEM 5.2

Develop a scientific program to deal with the nonlinear heat equation (5.4.3) on a slab with Neumann conditions in $y = 0$.

PROBLEM 5.3

Show how the solution method proposed in Section 5.4 can be technically generalized for the two-dimensional heat equation (5.5.1) with Neumann-type boundary conditions.

PROBLEM 5.4

Show how the solution method proposed in Section 5.4 can be technically generalized for the two-dimensional heat equation (5.5.1) when the diffusion coefficient K depends on the dependent variable u.

PROBLEM 5.5

Develop a scientific program to deal with the two-dimensional heat equation (5.5.1) over a circle domain using both Dirichlet and Neumann boundary conditions.

PROBLEM 5.6

Develop a scientific program to deal with the two-dimensional heat equation (5.5.1) over a domain as shown in the example of Figure 5.7.1.

PROBLEM 5.7

Develop a scientific program to deal with the Schnakenberg model with the initial and boundary conditions described in Section 5.6. Discuss the influence of the number of the collocation points.

PROBLEM 5.8

Develop a scientific program to deal with the Schnakenberg initial boundary problem (5.5.8)–(5.5.12) on a rectangular domain and show the influence of the aspect ratio.

PROBLEM 5.9

Write in detail the consistency conditions that have to be imposed on the collocation points belonging to the common boundaries of each subdomain D_k in the case of complex domain, as shown in the example of Figure 5.7.1.

PROBLEM 5.10

Develop a scientific program to deal with the heat equation (5.5.1), with Dirichlet boundary conditions, in the case of complex domain, as shown in the example of Figure 5.7.1.

6

Additional Mathematical Tools for Nonlinear Problems

6.1 Introduction

The initial-boundary value problems of the preceding chapters have been stated with linear boundary conditions. Nonlinearities have always been localized in the partial differential equations and not on the boundary conditions. However, the statements of boundary conditions, in several technological applications, are generated by measurements which cannot lead to a direct knowledge of the dependent variable and/or its space gradients. In some cases, measurements lead to nonlinear functions of the dependent variable at *the boundaries of the space variable domain.* In other cases, boundary measurements are not possible, and have to be substituted by additional information on the solution of the mathematical problem.

The term ***inverse problem*** is used to classify mathematical problems characterized by the above qualitative descriptions. Some of these problems may be ***ill posed*** as critically analyzed in Chapter 4 of the book by Bellomo and Preziosi (1996). It is worth analyzing how generalized collocation methods can be properly developed to deal with these types of problems due to their importance in applied sciences. The first part of this chapter deals with the above generalizations for models in one space dimension. Subsequently, the second part develops a critical analysis of the contents of this book. The contents are developed through eight more sections.

• Section 6.2 presents a *technical description of some ill-posed mathematical problems* which are dealt with later in this chapter. The statement refers both to problems with nonlinear boundary conditions and problems with lacunary conditions (boundary or source terms), which are replaced by additional measurements of the solution in interior points of the space domain.

• Section 6.3 concerns the *solution of problems with nonlinear boundary conditions.* The application of the method is technically shown for a *traffic flow model.* The model describes the evolution in time and space of the density of vehicles, while boundary conditions are delivered by measurements on the flux of vehicles that is given by a nonlinear combination of the Dirichlet and Neumann boundary conditions.

• Section 6.4 discusses the *solution of problems with missing boundary conditions.* The missing data are replaced by the measurement of the dependent variable in an inner zone of the domain of the space variable. The application of the method is technically shown for a *nonlinear diffusion model.*

• Section 6.5 concerns the *solution of problems with given boundary conditions, but with an unknown source term.* Such an unknown source term has to be identified by exploiting the additional information given by measurement of the dependent variable in an inner zone of the domain of the space variable. Subsequently, the method is applied to the solution of a *hydrodynamic model of pollution diffusion* introduced by Revelli and Ridolfi (2005).

• Section 6.6 shows how collocation methods can be applied to the solution of integro-differential equations arising in population dynamics. It is also shown how the method can operate in the case of ill-posed problems.

• Section 6.7 briefly discusses the use of orthogonal polynomials, which ensure a spectral approximation, as an alternative to the use of polynomials which identify precisely the interpolated function in the collocation points.

• Section 6.8 develops a *critical analysis* of the contents of this chapter and of the whole book referred also to a comparison with spectral methods. Advantages and drawbacks of the method are identified, and some research perspectives are outlined.

• Section 6.9 deals, as in the preceding chapters, with *problems brought to the attention of the reader.* Some of these problems are more difficult than those of the previous chapters. The reader can possibly consider them as perspectives for research programs in the field of engineering sciences.

Clearly, the selection of topics in this chapter is essentially due to a personal bias of the authors; additional subjects can be identified and solved after further developments of the method. The aim is to show how generalized collocation methods can be properly developed to the solution of a large variety of problems of interest in applied sciences.

6.2 On the Statement of Some Ill-Posed Problems

As we have seen in the previous chapters, mathematical models generate mathematical problems (in our case initial-boundary value problems), which are ***well formulated*** if the problem is stated with all initial and boundary conditions necessary for its solution. Problems which are not well formulated are said to be ***ill formulated.*** Well-formulated problems may also be ***well posed*** when the structure of the problem ensures that the solution exists, is unique, and depends continuously upon the data of the problem.

Improperly stated problems are ***underspecified*** if the information on the initial data and/or boundary conditions needed for its solution is incomplete, while the problem is ***overspecified*** if further information is given in addition to the initial and/or boundary conditions needed for its solution. The term ***inverse problem*** is used to assess initial-boundary value problems where some information on the initial and/or boundary conditions needed for its solution and/or on the parameters that characterize the model are missing and are replaced by suitable information on the solution of the mathematical problem.

Underspecified and ***overspecified*** problems are certainly ill posed. However, it is sometimes necessary, in engineering applications, to deal with them, whenever the mathematical formulation of the problem is imposed by particular practical situations. An ***underspecified*** problem occurs when not all initial and/or boundary conditions needed for its solution can be practically measured. On the other hand, ***overspecified*** problems occur when more measurements than those needed for their solution are given. The solution of a problem should show that the prediction of the model is close to the experimental additional measurements.

However, although a problem is properly stated, it can still be ill posed. This is the case, for instance, for nonlinear boundary conditions, which need to be assigned when measurements at the boundary involve nonlinear combinations of the Dirichlet and/or Neumann boundary conditions. This type of problem is generated, for instance, in the statement of initial-boundary value problems for the traffic flow models presented in Chapter 1, see Example 1.2.5, when measurements at the boundaries provide information on the flow rather than on the density of vehicles. Flow measurements, in this case, are relatively more precise than those related to the density. A specific application is proposed in the example which follows.

Example 6.2.1
Vehicular Traffic Models

Consider the traffic flow model proposed in Example 1.2.5:

$$\frac{\partial u}{\partial t} + (1-2u)\frac{\partial u}{\partial x} = \eta u^2(1-u)\frac{\partial^2 u}{\partial x^2} + \eta u(2-3u)\left(\frac{\partial u}{\partial x}\right)^2, \tag{6.2.1}$$

where $u = u(t,x)$ is the density referred to the maximum density of vehicles corresponding to a bumper-to-bumper traffic jam; t and x are, respectively, the dimensionless time and space variables referred to a characteristic time T and to the length of the road, where T is the time necessary to cover the whole road length at the maximum mean velocity; and η is a positive parameter related to the reactivity of the driver to the change of density.

The model is such that the flow q is referred to the local density and density gradients by the following relation:

$$q = 1 - u\left[1 + \eta(1-u)\frac{\partial u}{\partial x}\right] \cdot \tag{6.2.2}$$

The mathematical statement of the problem is obtained linking the above model with suitable initial conditions $\varphi(x) = u(0,x)$, and boundary conditions for the flow

$$\begin{cases} \psi(t) = h(\alpha(t),\gamma(t)) = 1 - \alpha(t)\left[1 + \eta(1-\alpha(t))\gamma(t)\right], \\ \xi(t) = k(\beta(t),\delta(t)) = 1 - \beta(t)\left[1 + \eta(1-\beta(t))\delta(t)\right], \end{cases} \tag{6.2.3}$$

where ψ and ξ are given functions of time corresponding to the inlet and outlet flows; α and β denote the Dirichlet boundary conditions

$$\alpha(t) = u(t,0), \qquad \text{and} \qquad \beta(t) = u(t,1), \tag{6.2.4}$$

while Neumann boundary conditions are identified by

$$\gamma(t) = \frac{\partial u}{\partial x}(t,0), \qquad \text{and} \qquad \delta(t) = \frac{\partial u}{\partial x}(t,1). \tag{6.2.5}$$

Writing the same model as a system of two equations, where the flow is used as an additional dependent variable, can be useful to test computational solutions. The model is as follows:

$$
\begin{cases}
\dfrac{\partial u}{\partial t} = -\dfrac{\partial q}{\partial x}\,, \\[2ex]
\dfrac{\partial q}{\partial t} = \left[(1-2u) + \eta u(2-3u)\dfrac{\partial q}{\partial x}\right]\dfrac{\partial q}{\partial x} + \eta u(1-u)\dfrac{\partial^2 q}{\partial x^2}\,.
\end{cases}
\tag{6.2.6}
$$

This problem is ill posed considering that Eqs. (6.2.3) do not allow us to uniquely identify the boundary conditions for the dependent variable. However, it is properly stated with the correct number of conditions needed for the solution.

□

An interesting class of problems with ***underspecified*** conditions can be stated if we add to the missing information suitable additional data related to the solution of the problem; these data have to be obtained by direct measurements. Some simple examples can be referred to the nonlinear heat equation in one space dimension as introduced in Chapter 1, Example 1.2.1,

$$
\frac{\partial u}{\partial t} = \frac{\partial}{\partial x}\left(\kappa(u)\frac{\partial u}{\partial x}\right), \tag{6.2.7}
$$

where $u = u(t,x)$ is the temperature for $t \geq 0$ and $x \in [0,1]$, and $\kappa = \kappa(u)$ is the heat conductivity, which depends on u.

The Dirichlet problem for this model is well formulated with initial condition

$$
\varphi(x) = u(0,x)\,, \tag{6.2.8}
$$

and boundary conditions (6.2.4), while the corresponding Neumann problem requires boundary conditions (6.2.5). On the other hand, the problem is ***underspecified*** if one of the two boundary conditions is not available. The problem becomes ***overspecified*** if, in addition to the above conditions (6.2.4) or (6.2.5), also the heat fluxes

$$
-\kappa\big(u(t,0)\big)\frac{\partial u}{\partial x}(t,0) \tag{6.2.9a}
$$

and/or

$$
-\kappa\big(u(t,1)\big)\frac{\partial u}{\partial x}(t,1) \tag{6.2.9b}
$$

are prescribed.

Underspecified problems may have the *missing boundary condition* replaced by the solution of the problem *in an interior point* of the space domain. Therefore, the problem can even be classified as an ***inverse type problem***. We can again use the heat diffusion model as an example.

Example 6.2.2
Inverse Problem for Nonlinear Diffusion Models—I

Consider the initial-boundary value problem Eq. (6.2.7) with the proper initial condition (6.2.8) and with only one of the two boundary conditions (6.2.4) needed for its solution. One can assume, for instance, that

$$\alpha(t) = u(t, 0)\,,$$

is prescribed while the value of $u(t, 1)$ is missing.

Consider now the case where the missing information is replaced by the solution to the initial-boundary value problem in the point $x = x^*$. That is,

$$u(t, x^*) = u^*(t)\,, \qquad \forall\, t \geq 0\,, \quad x^* \in]0, 1[\,. \tag{6.2.10}$$

The problem with ***unspecified boundary condition*** consists in determining the unknown boundary condition and the solution to the initial-boundary value problem.

□

According to the above reasoning, the following additional inverse problem is stated.

Example 6.2.3
Inverse Problem for Nonlinear Diffusion Models—II

The mathematical model is characterized by the addition of a well-localized source term and can be formally written as follows:

$$\frac{\partial u}{\partial t} = \frac{\partial}{\partial x}\left(\kappa(u)\frac{\partial u}{\partial x}\right) + s(t)\,\delta(x - x_s)\,, \tag{6.2.11}$$

where δ is the Dirac delta function and $0 < x_s < 1$ is given. The mathematical problem, with ***unspecified source term***, is stated with the proper initial and boundary conditions, say (6.2.8) and (6.2.4). The time evolution of the source term $s(t)$ is not known, while the solution (6.2.10) to the initial-boundary value problem is given for any $t \geq 0$ in some $x^* \in]0, 1[$. The problem consists in solving the initial-boundary value problem (6.2.11) and, in particular, in computing the source term $s(t)$.

□

It can be observed that the preceding problems are characterized by additional information analogous to the missing information. In other words, a function depending on time is replaced by information given as a function of time. These problems are characterized by instability features, as well documented in the book by Back, Blackwell and Saint Claire (1985), which is devoted to the analysis of various inverse problems for the heat equation; see also Lavrent'ev, Reznitskaya, and Yakhno (1985) and Colton, Ewing, and Rundell (1990).

Suitable developments of the generalized collocation methods can efficiently solve such problems, as we shall see in the following sections.

6.3 Problems with Nonlinear Boundary Conditions

Problems in applied sciences may require, as already mentioned, nonlinear boundary conditions. The solution method proposed below is based on the idea of extracting from the nonlinear boundary conditions a nonlinear evolution equation for the Dirichlet boundary conditions

$$\alpha(t) = u(t, 0)\,, \qquad \text{and} \qquad \beta(t) = u(t, 1)\,,$$

where the dimensionless dependent variable

$$u = u(t, x) : \quad \mathbb{R}_+ \times [0, 1] \to [0, 1]$$

describes, in the mathematical model, the state of a real physical system.

Within a general abstract framework, nonlinear boundary conditions can be given as follows:

$$\begin{cases} \psi(t) = h(\alpha(t), \gamma(t))\,, \\ \xi(t) = k(\beta(t), \delta(t))\,, \end{cases} \tag{6.3.1}$$

where ψ and ξ are given smooth functions of time, and h and k are given smooth functions of their arguments $\alpha, \beta, \gamma, \delta$, while

$$\gamma(t) = \frac{\partial u}{\partial x}(t, 0)\,, \qquad \text{and} \qquad \delta(t) = \frac{\partial u}{\partial x}(t, 1)\,.$$

The derivation of ψ and ξ with respect to time yields

$$\frac{d\psi}{dt} = h_\alpha(\alpha, \gamma)\frac{d\alpha}{dt} + h_\gamma(\alpha, \gamma)\frac{d\gamma}{dt} \tag{6.3.2}$$

and

$$\frac{d\xi}{dt} = k_\beta(\beta, \delta)\frac{d\beta}{dt} + k_\delta(\beta, \delta)\frac{d\delta}{dt}\,, \tag{6.3.3}$$

where subscripts denote partial derivatives:

$$h_\alpha = \frac{dh}{d\alpha}\,, \quad h_\gamma = \frac{dh}{d\gamma}\,, \quad k_\beta = \frac{dk}{d\beta}\,, \quad k_\delta = \frac{dk}{d\delta}\,.$$

The terms γ and δ can be approximated by the following Lagrange interpolation:

$$\gamma(t) = a_{11}u_1(t) + \sum_{j=2}^{n-1} a_{j1}u_j(t) + a_{n1}u_n(t)\,, \tag{6.3.4}$$

and

$$\delta(t) = a_{1n}u_1(t) + \sum_{j=2}^{n-1} a_{jn}u_j(t) + a_{nn}u_n(t)\,. \tag{6.3.5}$$

Therefore, recalling that $\alpha = u_1$ and $\beta = u_n$, the derivation with respect to time yields

$$\frac{d\gamma}{dt} = a_{11}\frac{d\alpha}{dt} + \sum_{j=2}^{n-1} a_{j1}\frac{du_j}{dt} + a_{n1}\frac{d\beta}{dt} \tag{6.3.6}$$

and

$$\frac{d\delta}{dt} = a_{1n}\frac{d\alpha}{dt} + \sum_{j=2}^{n-1} a_{jn}\frac{du_j}{dt} + a_{nn}\frac{d\beta}{dt}\,. \tag{6.3.7}$$

Let us now denote by $\mathbf{u}^*$ the vector: $\{u_2\,, \dots\,, u_{n-1}\}$; then, substituting (6.3.6) and (6.3.7) into (6.3.2) and (6.3.3) yields

$$\begin{cases} [h_\alpha(\alpha,\beta,\mathbf{u}^*) + a_{11}h_\gamma(\alpha,\beta,\mathbf{u}^*)]\dfrac{d\alpha}{dt} + a_{n1}h_\gamma(\alpha,\beta,\mathbf{u}^*)\dfrac{d\beta}{dt} \\ \qquad = \dfrac{d\psi}{dt} - h_\gamma(\alpha,\beta,\mathbf{u}^*)\displaystyle\sum_{j=2}^{n-1} a_{j1}\frac{du_j}{dt}\,, \\ \\ a_{1n}k_\delta(\alpha,\beta,\mathbf{u}^*)\dfrac{d\alpha}{dt} + [k_\beta(\alpha,\beta,\mathbf{u}^*) + a_{nn}k_\delta(\alpha,\beta,\mathbf{u}^*)]\dfrac{d\beta}{dt} \\ \qquad = \dfrac{d\xi}{dt} - k_\delta(\alpha,\beta,\mathbf{u}^*)\displaystyle\sum_{j=2}^{n-1} a_{jn}\frac{du_j}{dt}\,. \end{cases} \tag{6.3.8}$$

The algebraic system (6.3.8) can be solved with respect to the ***time derivatives*** of α and β which occur in the first and last equations. The scientific program will then be an architecture of two programs suitable for dealing with two different problems: solution of the algebraic system and integration of the differential system. The following example shows how this method is well suited to deal with the model reported in Example 6.2.1.

Example 6.3.1
Vehicular Traffic Models

Consider the model proposed in Example 6.2.1 which we write again as follows:

$$\frac{\partial u}{\partial t} + (1-2u)\frac{\partial u}{\partial x} = \eta u^2(1-u)\frac{\partial^2 u}{\partial x^2} + \eta u(2-3u)\left(\frac{\partial u}{\partial x}\right)^2 , \tag{6.3.9}$$

complemented with the boundary conditions

$$\begin{cases} \psi(t) = h(\alpha(t),\gamma(t)) = 1 - \alpha(t)\left[1+\eta(1-\alpha(t))\gamma(t)\right] , \\ \xi(t) = k(\beta(t),\delta(t)) = 1 - \beta(t)\left[1+\eta(1-\beta(t))\delta(t)\right] , \end{cases} \tag{6.3.10}$$

and initial condition

$$u(x, t=0) = u_0(x) = 4x^2(x-1)^2 . \tag{6.3.11}$$

Simulations have been performed with the particular choice of $\eta = 0.1$, and with constantly null inlet flow $\psi = 0$, and outlet flow given by $\xi(t) = 0$.

□

Figure 6.3.1.a shows the simulation obtained using Lagrange polynomials with 41 nodes. Some irregularities are evident, testifying to the complexity of the problem. The numerical result improves if the number of nodes, n, is increased; specifically, Figures 6.3.1.b–c show the behavior of $u(x,t)$ when the point number is 101 and 151, respectively. In particular, in the latter case, the numerical simulation is very regular and smooth. The simulations shown in Figures 6.3.1.a–c are obtained with the help of the program ***DiffNL*** (Appendix, Section A.17).

In general, the above procedure drastically simplifies when only one of the boundary conditions is nonlinear, say $\psi(t) = h(\alpha(t),\gamma(t))$, with given $\beta(t)$, or $\xi(t) = k(\beta(t),\delta(t))$ with given $\alpha(t)$. In this case the reference equations are, respectively, Eq. (6.3.2) or Eq. (6.3.3).

In this case the calculations are simplified as only one of the two evolution equations needs to be used, and each equation of (6.3.8) can be dealt with separately from the other. In detail:

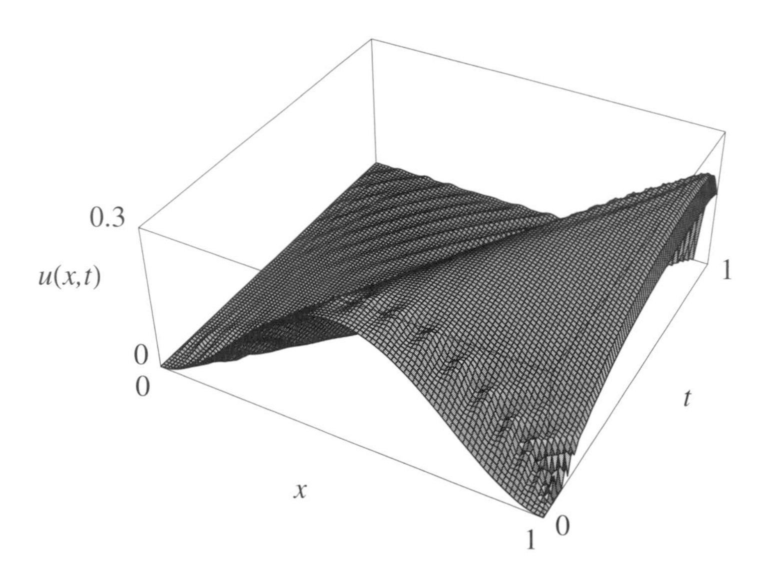

Figure 6.3.1.a - Simulation with 41 nodes.

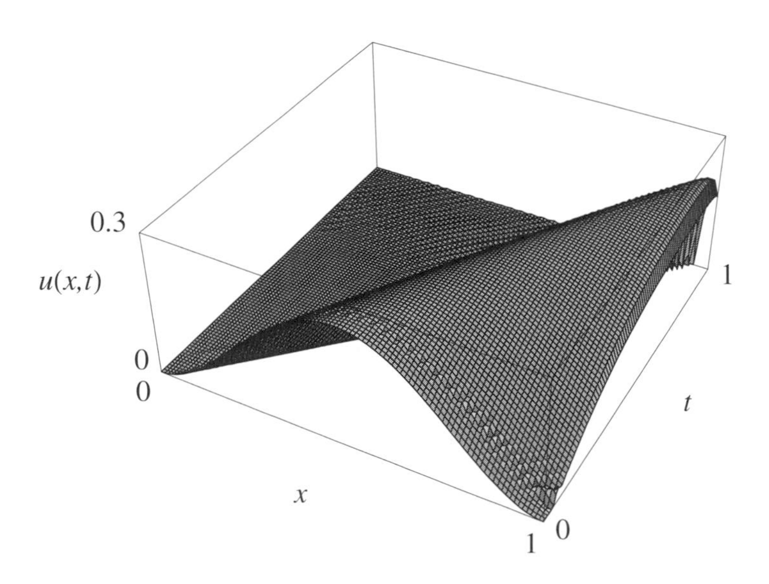

Figure 6.3.1.b - Simulation with 101 nodes.

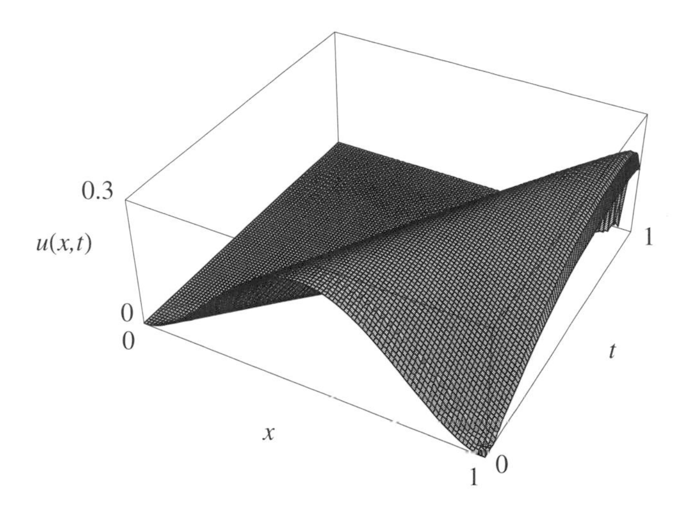

Figure 6.3.1.c - Simulation with 151 nodes.

$$
\begin{aligned}
\frac{d\alpha}{dt} &= \frac{1}{[h_\alpha(\alpha,\beta,\mathbf{u}^*) + a_{11}h_\gamma(\alpha,\beta,\mathbf{u}^*)]}\,\frac{d\psi}{dt} \\
&- \frac{h_\gamma(\alpha,\beta,\mathbf{u}^*)}{[h_\alpha(\alpha,\beta,\mathbf{u}^*) + a_{11}h_\gamma(\alpha,\beta,\mathbf{u}^*)]}\,\sum_{j=2}^{n-1} a_{j1}\,\frac{du_j}{dt} \\
&- \frac{a_{n1}h_\gamma(\alpha,\beta,\mathbf{u}^*)}{[h_\alpha(\alpha,\beta,\mathbf{u}^*) + a_{11}h_\gamma(\alpha,\beta,\mathbf{u}^*)]}\,\frac{d\beta}{dt}\,, \qquad (6.3.12)
\end{aligned}
$$

or

$$
\begin{aligned}
\frac{d\beta}{dt} &= \frac{1}{[k_\beta(\alpha,\beta,\mathbf{u}^*) + a_{nn}k_\delta(\alpha,\beta,\mathbf{u}^*)]}\,\frac{d\xi}{dt} \\
&- \frac{k_\delta(\alpha,\beta,\mathbf{u}^*)}{[k_\beta(\alpha,\beta,\mathbf{u}^*) + a_{nn}k_\delta(\alpha,\beta,\mathbf{u}^*)]}\,\sum_{j=2}^{n-1} a_{jn}\,\frac{du_j}{dt} \\
&- \frac{a_{1n}k_\delta(\alpha,\beta,\mathbf{u}^*)}{[k_\beta(\alpha,\beta,\mathbf{u}^*) + a_{nn}k_\delta(\alpha,\beta,\mathbf{u}^*)]}\,\frac{d\alpha}{dt}\,. \qquad (6.3.13)
\end{aligned}
$$

The solution can be compared with the one obtained for model (6.2.6) for a Dirichlet problem corresponding to the one with nonlinear boundary conditions.

6.4 Problems with Unspecified Boundary Conditions

This section deals with the solution of the problem described in Example 6.2.2, where the boundary condition which is missing is replaced by $u(t, x^*)$ for an interior point x^* of the space domain. The description of the solution techniques to the problem can be referred to the following solution schemes.

Scheme 1: Consider the initial-boundary value problem related to the diffusion equation (6.2.7) with two-point boundary conditions (6.2.4), for $t \geq 0$ and initial conditions (6.2.8). According to Chapter 4, this problem can be solved by introducing a suitable *space collocation*

$$I_x = \{x_1 = 0, \ldots, x_{n-1}, x_n = 1\}. \tag{6.4.1}$$

The solution method is based on the approximation of the solution $u = u(t, x)$ by its interpolation

$$u(t, x) \cong u^n(t, x) = \sum_{j=1}^{n} u_j(t) L_j(x), \tag{6.4.2}$$

where L_j denote the Lagrange polynomials corresponding to the above discretization, and $u_j = u(t, x_j)$, for $j = 1, \ldots, n$. Interpolation (6.4.2) allows us to approximate the space derivatives in the nodal points as shown in Chapter 2.

The time evolution of the values of the dependent variable in the nodes $u_i(t) = u(t, x_i)$ are then obtained by substituting the expression of u delivered by the interpolation (6.4.2) and of the related space derivatives into Eq. (6.2.7) where boundary conditions have to be enforced. This yields a system of nonlinear ordinary differential equations of the type

$$\frac{du_i}{dt} = \kappa(u_i) \sum_{j=1}^{n} b_{ij} u_j + \kappa_u(u_i) \left(\sum_{j=1}^{n} a_{ij} u_j \right)^2, \tag{6.4.3}$$

for $i = 2, \ldots, n-1$, with $\kappa_u = \dfrac{\partial \kappa}{\partial u}$, and the first and last equation of the system (corresponding respectively to $i = 0$ and $i = n$) are replaced by

$$u_0(t) = \alpha(t), \qquad u_n(t) = \beta(t). \tag{6.4.4}$$

Recall that here we use the shorthand notation

$$b_{ij} = a_{ij}^{(2)} \quad \text{and} \quad a_{ij} = a_{ij}^{(1)}, \quad i,j = 1,\ldots,n\,,$$

where the technical expression of $a_{ij}^{(r)}$ is reported in Chapter 2, Eq. (2.2.11). This scheme consists of a suitable collocation in space followed by integration in time.

Scheme 2: Let us consider now the initial-boundary value problem for the heat equation (6.2.7) with initial condition (6.2.8) and boundary conditions

$$u(t,0) = \alpha(t)\,, \qquad \frac{\partial u}{\partial x}(t,0) = \gamma(t)\,. \tag{6.4.5}$$

The evolution equation can be rewritten as a first-order system, with respect to the space derivatives, as follows:

$$\begin{cases} \dfrac{\partial u}{\partial x} = \dfrac{1}{\kappa(u)}\,w\,, \\ \dfrac{\partial w}{\partial x} = \dfrac{\partial u}{\partial t}\,. \end{cases} \tag{6.4.6}$$

The solution method proceeds in a fashion similar to ***Scheme 1***, but now introducing, instead of the space collocation I_x, a suitable *time collocation*:

$$I_t = \{t_1 = 0, \ldots, t_{m-1}, t_m = 1\}\,.$$

Therefore, the state variable is approximated as follows:

$$u(t,x) \cong u^m(t,x) \stackrel{\text{def}}{=} \sum_{j=1}^{m} L_j(t)u_j(x)\,, \tag{6.4.7}$$

where $u_j(x) = u(t_j,x)\,, \quad j = 1,\ldots,m$.

Similar to ***Scheme 1***, the time derivative in the nodal points can then be approximated, with the obvious meaning of symbols, as follows:

$$\frac{\partial u}{\partial t}(t_i,x) \cong \sum_{j=1}^{m} a_{ij}u_j(x)\,. \tag{6.4.8}$$

Substitution of the expression (6.4.8) into Eq. (6.4.6) yields the following system of ordinary differential equations:

$$\begin{cases} \dfrac{du_i}{dx} = \dfrac{1}{\kappa(u_i)} w_i\,, \\[2ex] \dfrac{dw_i}{dx} = \displaystyle\sum_{j=1}^{m} a_{ij} u_j\,, \end{cases} \tag{6.4.9}$$

which defines the evolution, in space, of u and w corresponding to the nodal points in time. The initial conditions are given at $x = 0$:

$$u_i(x=0) = \alpha(t_i)\,, \qquad w_i(x=0) = \kappa(\alpha(t_i))\,\gamma(t_i)\,. \tag{6.4.10}$$

In other words, the boundary conditions of the original problem act as initial conditions of the new problem, while the initial condition $u(0,x) = \varphi(x)$ acts as a boundary condition. In conclusion, this scheme is based on a ***collocation in time and integration in space.***

Consider now the mathematical problem stated in Example 6.2.2. The solution of this problem needs the ***decomposition of the domain*** $\mathcal{D} = [0,1]$ of the space variable into two subdomains

$$\mathcal{D} = \mathcal{D}_1 \cup \mathcal{D}_2\,, \tag{6.4.11}$$

where

$$\mathcal{D}_1 = [0, x^*] \quad \text{and} \quad \mathcal{D}_2 = [x^*, 1]\,.$$

In order to avoid ambiguities in the application of solution algorithms, the dependent variable in $\mathcal{D}_1$ and $\mathcal{D}_2$ will be denoted, respectively, by $u^{(1)}$ and $u^{(2)}$. The initial-boundary value problem in $\mathcal{D}_1$ for Eq. (6.4.4) is solved by ***Scheme 1***. Therefore, $\varphi(x)$ is the initial condition, whereas $\alpha(t)$ and $u^*(t)$ act as boundary conditions.

The problem in $\mathcal{D}_2$ is solved, subsequently, by ***Scheme 2***. In this case the role of initial condition is played by $u^*(t)$ and the value of the space derivative of the temperature in $x = x^*$ is obtained by the solution over $\mathcal{D}_1$ as

$$\begin{aligned} w^{(2)}(t, x^*) &= \kappa\big(u^{(1)}(t, x^*)\big) \frac{\partial u^{(1)}}{\partial x}(t, x^*) \\ &= \kappa\big(u^{(1)}(t, x^*)\big) \sum_{j=1}^{n} a_{nj} u_j^{(1)}(t)\,. \end{aligned} \tag{6.4.12}$$

The specific problem developed in Section 6.6 will show a practical application of this solution scheme for a model different from the one used in this section.

6.5 On the Identification of Source Terms

Consider now the solution of the initial-boundary value problem for the heat equation with source term (6.2.11) stated in ***Example 6.2.3***. Specifically, consider a problem with ***unspecified source term*** stated with the proper initial and boundary conditions, while the time evolution of the source term $s(t)$, localized in x_s, is not known. On the other hand, the solution

$$u(t, x^*) = u^*(t) \tag{6.5.1}$$

to the initial-boundary value problem is given for any $t \geq 0$ and for some $x^* \in]0, 1[$. As already stated, the problem consists in solving the initial-boundary value problem and, in particular, computing the source term $s(t)$.

This problem can be solved by decomposing the original problem into several, properly linked, direct problems which are all well formulated. Therefore, the solution method consists in using, for each of these direct problems, one of the Schemes (1 or 2) described in Section 6.4.

Precisely, the domain $\mathcal{D}$ of the space variable is decomposed into three subdomains:

$$\mathcal{D} = \mathcal{D}_1 \cup \mathcal{D}_2 \cup \mathcal{D}_3 \,, \tag{6.5.2}$$

where

$$\mathcal{D}_1 = [0, x^*] \,, \qquad \mathcal{D}_2 = [x^*, x_s] \,, \qquad \mathcal{D}_3 = [x_s, 1] \,.$$

Moreover, in accordance with the notation of Section 6.4, the dependent variable u, in each subdomain $\mathcal{D}_i$, will be denoted by $u^{(i)}$, $i = 1, 2, 3$.

The problem in $\mathcal{D}_1$ and $\mathcal{D}_2$ is solved using the solution method developed in Section 6.4. As a consequence, the solution in x_s,

$$u^{(2)}(t, x_s) = u_s(t), \tag{6.5.3}$$

is computed by the method developed above.

The problem in $\mathcal{D}_3$ is then well formulated with the proper initial and boundary conditions $\varphi(x)$, $u_s(t)$, and $\beta(t)$. Therefore, this problem can be solved by ***Scheme 1***, and the source term is identified by the change of slope in the dependent variable:

$$s(t) = -\kappa\big(u_s(t)\big)\left(\frac{\partial u}{\partial x}^{(3)}(t, x_s) - \frac{\partial u}{\partial x}^{(3)}(t, x_s)\right). \tag{6.5.4}$$

We stress that the space derivative of $u^{(2)}$ is known, by the integration by ***Scheme 2***, at the time collocation points t_i,

$$\frac{\partial u}{\partial x}^{(2)}(t_i, x_s) = \frac{1}{\kappa(u_s(t_i))} w^{(2)}(t_i). \tag{6.5.5}$$

Therefore, its value for any t is obtained by interpolation,

$$\frac{\partial u}{\partial x}^{(2)}(t, x_s) = \sum_{i=1}^{m} \frac{1}{\kappa(u_s(t_i))} w^{(2)}(t_i) L_i(t). \tag{6.5.6}$$

On the other hand,

$$\frac{\partial u}{\partial x}^{(3)}(t, x_s) = \sum_{j=1}^{n} L_j(x) u_j^{(3)}(t) \tag{6.5.7}$$

is available essentially at any t, since in $\mathcal{D}_3$ we are integrating in time.

These problems refer to systems defined over space domains which are fixed in time. However, several problems of interest in applied science are defined over domains which evolve in time. For instance, this is the case for diffusion problems with phase transition at the boundary, where the localization of the boundary is an unknown variable, whose evolution depends on the local slope of the temperature profiles and on the latent heat of the phase transition.

If the temperatures in the domains D_1 and D_2 are denoted, respectively, by $u = u(t, x)$ and $v = v(t, x)$, the statement of the mathematical problem is as follows:

$$\begin{cases} \dfrac{\partial u}{\partial t} = \dfrac{\partial}{\partial x}\left(\kappa(u)\dfrac{\partial u}{\partial x}\right), \\[2ex] \dfrac{\partial v}{\partial t} = \dfrac{\partial}{\partial x}\left(\kappa(v)\dfrac{\partial v}{\partial x}\right) \\[2ex] \dfrac{dz}{dt} = L(w)\left(\kappa(u)\dfrac{\partial u}{\partial x}|_{x=z} - \kappa(v)\dfrac{\partial v}{\partial x}|_{x=z}\right), \end{cases} \tag{6.5.8}$$

where $z = z(t)$ is the coordinate of the moving boundary, and w is the temperature of the phase transition.

It can be briefly shown how the methods proposed in this chapter can be technically generalized to deal with this, problem. In fact, it is sufficient to scale the space variables with respect to the variable $z = z(t)$ to obtain a problem analogous to those we have seen above, where now z is an additional variable. The interested reader now has sufficient information to deal technically with this problem.

The preceding solution technique is applied here to a model proposed by Revelli and Ridolfi (2005) to describe the main transport processes that a chemical undergoes in a river. The model, reported in Chapter 1, Example 1.2.4, is rewritten here, in its dimensionless form, for completeness:

$$\begin{aligned}\frac{\partial u}{\partial t} &= [\varepsilon_1 f_1(x) - \varepsilon_2 \varphi_2(x)] \frac{\partial u}{\partial x} + \varepsilon_3 \varphi_3(x) \frac{\partial^2 u}{\partial x^2} \\ &\quad - \mu u^m - \varepsilon_4 \varphi_4(x) u + \eta s(t) q(x) ,\end{aligned} \tag{6.5.9}$$

where the dimensionless parameters ε_i with $i = 1, 2, 3, 4$ are defined as follows:

$$\begin{aligned}&\varepsilon_1 = \frac{a_K b_K T_c}{\ell^2}, \qquad \varepsilon_2 = \frac{a_v T_c}{\ell}, \qquad \varepsilon_3 = \frac{a_K T_c}{\ell^2} = \frac{\varepsilon_1}{b_k} , \\ &\qquad\qquad \varepsilon_4 = \frac{a_v b_v T_c}{\ell} = \varepsilon_2 b_v ,\end{aligned} \tag{6.5.10}$$

while the functions of the space variable turn out to be

$$\begin{aligned}\varphi_1(x) &= (1 + x)^{b_K - 1}, \qquad \varphi_2(x) = (1 + x)^{b_v} , \\ \varphi_3(x) &= (1 + x)^{b_K}, \qquad \varphi_4(x) = (1 + x)^{b_v - 1} ,\end{aligned}$$

and the constants μ and ν are

$$\mu = \lambda C_M^{m-1} T_c , \qquad \text{and} \qquad \eta = c T_c C_M^{-1} .$$

The constants a_v, b_v, a_K, b_K, λ, m, and c have been defined in Example 1.3.2.

The source term $s(t)$ is concentrated at the position $x = x_s$, with $0 < x_s < 1$, i.e.,

$$q(x) = \delta(x - x_s) ,$$

while the concentrations of pollutant in $0 < x_* < x_s$ and $x_s < x^* < 1$ are denoted, respectively, by $u_*(t) = u(t, x_*)$ and $u^*(t) = u(t, x^*)$.

We now discuss the following three problems.

Problem 6.5.1. Compute $u(t,x)$ with given $\alpha(t)$, $\beta(t)$, $\varphi(x)$, and $s(t)$, where α and β are the Dirichlet-type boundary conditions and φ is the initial condition.

Problem 6.5.2. Compute $s(t)$ and $u(t,x)$ with given $\alpha(t)$, $\beta(t)$, $u_*(t)$ (or $u^*(t)$), and $\varphi(x)$.

Problem 6.5.3. Compute $\beta(t)$ and $u(t,x)$ with given $\alpha(t)$, $s(t)$, $\varphi(x)$, and $u^*(t)$.

Problem 6.5.1 is a direct problem which requires a suitable domain decomposition scheme; then it can be solved by the collocation method described in Chapter 3. Problem 6.5.2 can be classified as an inverse problem, which aims to compute the unknown temporal evolution of the source when its spatial position, the concentration in a downstream (or upstream) point, and the boundary conditions are known. In contrast, Problem 6.5.3 deals with the information on the downstream boundary condition given by the source term.

Let us now consider Problem 6.5.1 and introduce the following domain decomposition:

$$\mathcal{D} = [0,1] = \mathcal{D}_1 \cup \mathcal{D}_2 = [0, x_s] \cup [x_s, 1] \tag{6.5.11}$$

with Chebychev space collocation with n_1 nodes in the subdomain $\mathcal{D}_1$ and n_2 nodes in $\mathcal{D}_2$.

Scheme 1 is used in both domains $\mathcal{D}_1$ and $\mathcal{D}_2$, and the dependent variable $u^{(r)} = u^{(r)}(t,x)$ is interpolated and approximated by means of the values $u_i^{(r)}(t) = u^{(r)}(t, x_i^{(r)})$ using Lagrange polynomials, i.e.,

$$u^{(r)}(t,x) \cong \sum_{j=1}^{n_r} L_j^{(r)}(x)\, u_j^{(r)}(t)\,,$$

where $r = 1, 2$. Here, the Lagrange polynomials $L_j^{(r)}(x)$ obviously depend on r through the number of interpolation nodes n_r, $r = 1, 2$.

This interpolation is used to approximate the partial derivatives of the variable u in the nodal points of the discretization in both domains:

$$\frac{\partial u}{\partial x}^{(r)}(t; x_i^{(r)}) \cong \sum_{j=1}^{n_r} \frac{dL_j^{(r)}}{dx}(x_i^{(r)}) u_j^{(r)}(t) = \sum_{j=1}^{n_r} a_{ji}^{(r)} u_j^{(r)}(t)\,,$$

and

$$\frac{\partial^2 u}{\partial x^2}^{(r)}(t; x_i^{(r)}) \cong \sum_{j=1}^{n_r} \frac{d^2 L_j^{(r)}}{dx^2}(x_i^{(r)}) u_j^{(r)}(t) = \sum_{j=1}^{n_r} b_{ji}^{(r)} u_j^{(r)}(t) ,$$

where, with the notation of Chapter 2,

$$a_{ij}^{(r)} = a_{ij}^{(1)}(n_r) , \quad \text{and} \quad b_{ji}^{(r)} = a_{ij}^{(2)}(n_r) .$$

The solution is obtained by integration of the following system of ordinary differential equations:

$$\begin{cases} \dfrac{du_i}{dt}^{(1)} = f_i^{(1)}(u_1^{(1)}, \dots, u_{n_1}^{(1)}) , \qquad i = 1, \dots, n_1 - 1 , \\ \dfrac{du_j}{dt}^{(2)} = f_j^{(2)}(u_1^{(2)}, \dots, u_{n_2}^{(2)}) , \qquad j = 2, \dots, n_2 , \\ \quad u_s \equiv u_{n_1}^{(1)} \equiv u_1^{(2)} = f^{(1,2)}(u_1^{(1)}, \dots, u_{n_1-1}^{(1)}, u_2^{(2)}, \dots, u_{n_2}^{(2)}) , \end{cases} \tag{6.5.12}$$

where for $i = 1$,

$$f_1^{(1)} = \frac{d\alpha}{dt} , \tag{6.5.13}$$

and for $i = 2, \dots, n_1 - 1$,

$$\begin{aligned} f_i^{(1)} = {} & [\varepsilon_1 \varphi_1(x_i^{(1)}) - \varepsilon_2 \varphi_2(x_i^{(1)})] \left[\sum_{j=1}^{n} a_{ji}^{(1)} u_j^{(1)} \right] \\ & + \varepsilon_3 \varphi_3(x_i^{(1)}) \left[\sum_{j=1}^{n} b_{ji}^{(1)} u_j^{(1)} \right] \\ & - \mu u_i^{(1)\, m} - \varepsilon_4 \varphi_4(x_i^{(1)})\, u_i^{(1)} . \end{aligned} \tag{6.5.14}$$

On the other hand, for $j = 2, \dots, n_2 - 1$, one has

$$\begin{aligned} f_j^{(2)} = {} & [\varepsilon_1 \varphi_1(x_j^{(2)}) - \varepsilon_2 \varphi_2(x_j^{(2)})] \left[\sum_{k=1}^{n_2} a_{kj}^{(2)} u_k^{(2)} \right] \\ & + \varepsilon_3 \varphi_3(x_j^{(2)}) \left[2 \sum_{k=1}^{n_2} b_{kj}^{(2)} u_k^{(2)} \right] \\ & - \mu u_j^{(2)\, m} - \varepsilon_4 \varphi_4(x_j^{(2)})\, u_j^{(2)} , \end{aligned} \tag{6.5.15}$$

and for $j = n_2$,

$$f_{n_2}^{(2)} = \frac{d\beta}{dt} \,. \tag{6.5.16}$$

Moreover, an additional condition is needed:

$$s(t) = \varepsilon_3 \left[\left.\frac{\partial u}{\partial x}\right|_{x=x_s^+} - \left.\frac{\partial u}{\partial x}\right|_{x=x_s^-} \right] , \tag{6.5.17}$$

which implies

$$\varepsilon_3 \left[\sum_{j=1}^{n_1-1} a_{n_1 j}^{(1)} u_j^{(1)} + a_{n_1 n_1}^{(1)} u_s - \sum_{j=2}^{n_2} a_{1j}^{(2)} u_j^{(2)} - a_{11}^{(2)} u_s \right] = s(t), \tag{6.5.18}$$

so that

$$f^{(1,2)} = u_s = \frac{1}{(a_{n_1 n_1}^{(1)} - a_{11}^{(2)})} \left\{ \frac{s(t)}{\varepsilon_3} + \sum_{j=2}^{n_2} a_{1j}^{(2)} u_j^{(2)} - \sum_{j=1}^{n_1-1} a_{n_1 j}^{(1)} u_j^{(1)} \right\} . \tag{6.5.19}$$

The above problem is solved with suitable initial conditions delivered by $\varphi(x)$ in both domains. Then, the values $u^{(r)}(t,x)$, $r = 1, 2$, are finally computed.

Consider now the solution to Problem 6.5.2, which requires the following domain decomposition:

$$\mathcal{D} = [0,1] = \mathcal{D}_1 \cup \mathcal{D}_2 \cup \mathcal{D}_3 = [0, x_*] \cup [x_*, x_s] \cup [x_s, 1] \tag{6.5.20}$$

and the use of Scheme 1 for the $\mathcal{D}_1$ and $\mathcal{D}_3$ domains and Scheme 2 for the $\mathcal{D}_2$ domain.

The dependent variable is interpolated and approximated, in $\mathcal{D}_1$ and $\mathcal{D}_3$, by Lagrange polynomials as in Problem 6.5.1. However, in $\mathcal{D}_2$ it is often more convenient to use an interpolation by sinc functions.

The solution scheme in domains $\mathcal{D}_1$ and $\mathcal{D}_3$ is the same as that of Problem 6.5.1. However, in the $\mathcal{D}_2$ domain it is necessary to introduce a new dependent variable, v, by changing the model into the system

$$\begin{cases} \dfrac{\partial u^{(2)}}{\partial x} = v^{(2)} \,, \\[2ex] \dfrac{\partial v^{(2)}}{\partial x} = \dfrac{1}{\varepsilon_3 \varphi_3} \dfrac{\partial u^{(2)}}{\partial t} - \dfrac{\varepsilon_1 \varphi_1 - \varepsilon_2 \varphi_2}{\varepsilon_3 \varphi_3} v^{(2)} \\[2ex] \qquad + \dfrac{\mu}{\varepsilon_3 \varphi_3} u^{(2)\, m} + \dfrac{\varepsilon_4 \varphi_4}{\varepsilon_3 \varphi_3} u^{(2)} \,. \end{cases} \tag{6.5.21}$$

The following system is obtained:

$$\begin{cases} \dfrac{du_i^{(2)}}{dx} = f_i^{(2)}(v_1^{(2)}, \dots, v_{n_2}^{(2)}), \\ \dfrac{dv_i^{(2)}}{dx} = g_i^{(2)}(u_1^{(2)}, \dots, u_{n_2}^{(2)}, v_1^{(2)}, \dots, v_{n_2}^{(2)}) \end{cases} \tag{6.5.22}$$

for $i = 1, \dots, n_2$, where

$$f_1^{(2)} = \gamma(x), \tag{6.5.23a}$$

$$g_1^{(2)} = \gamma_x(x), \tag{6.5.23b}$$

and

$$\begin{cases} f_i^{(2)} = v_i^{(2)}, \\ g_i^{(2)} = \dfrac{1}{\varepsilon_3\varphi_3(x)} \left[\displaystyle\sum_{j=1}^{n_2} a_{ji}^{(2)} u_j^{(2)} \right] - \dfrac{\varepsilon_1\varphi_1(x) - \varepsilon_2\varphi_2(x)}{\varepsilon_3\varphi_3(x)} v_i^{(2)} \\ \qquad + \dfrac{\varepsilon_3\varphi_3(x)}{\varepsilon_4\varphi_4(x)} u_i^{(2)} + \dfrac{\mu}{\varepsilon_3\varphi_3(x)} [u_i^{(2)}]^m, \end{cases} \tag{6.5.24}$$

for $i = 2, \dots, n_2$.

The initial conditions are given by

$$\begin{cases} u(t, x_*) = \displaystyle\sum_{j=1}^{n_1} L_j^{(1)}(x_*) u_j^{(1)}(t), \\ v(t, x_*) = \displaystyle\sum_{j=1}^{n_1} a_{jn_1}^{(1)}(x_*) u_j^{(1)}(t). \end{cases} \tag{6.5.25}$$

Finally, let us consider Problem 6.5.3. The solution method is similar to the one of Problem 6.5.2. The subdomains are the same and Scheme 1 is adopted for the $\mathcal{D}_1$ domain, while Scheme 2 is used for the $\mathcal{D}_2$ and $\mathcal{D}_3$ domains. In particular, Scheme 1 is used in $\mathcal{D}_1$ to compute $u_x^{(1)}(x_*, t)$. Then, Scheme 2 in $\mathcal{D}_2$ allows to obtain u_s, $u_x^{(2)}(x_s, t)$, and, by means of the joint condition (6.5.19), $u_x^{(3)}(x_s, t)$. Finally, by applying Scheme 2 in the domain $\mathcal{D}_3$, the boundary condition $\beta(t)$ is computed.

Remark 6.5.1. *The solution methods proposed for the inverse Problems 6.5.2 and 6.5.3 can also be applied when the measure, $u^*(t)$, is localized* *downstream of the source, that is $x^* > x_s$. In this case it is sufficient to define the following domain decomposition:*

$$\mathcal{D} = [0,1] = \mathcal{D}_1 \cup \mathcal{D}_2 \cup \mathcal{D}_3 = [0, x_s] \cup [x_s, x^*] \cup [x^*, 1]$$

and to apply the same steps proposed for Problem 6.5.2 or 6.5.3, respectively.

The application developed below shows how the method can provide computational solutions to these problems. The following values of the parameters are used:

$$i_b = 0.001\,, \qquad \ell = 33600\,[m]\,,$$

$$a_h = 1.274\,[m]\,, \qquad b_h = 0.74\,,$$

$$a_v = 0.285\,[ms^{-1}]\,, \qquad b_v = 0.19\,,$$

$$\varepsilon = 2.42 \cdot 10^{-5}\,, \qquad C_M = 2 \cdot 10^{-3}\,[\mathrm{kgm}^{-3}]\,.$$

Detailed calculations based on these constants yield

$$a_k = 0.845\,[\mathrm{m}^2\mathrm{s}^{-1}]\,, \qquad b_k = 1.11\,, \qquad T = 32400\,[\mathrm{s}]\,,$$

and

$$\lambda = 10^{-4}\,[\mathrm{Kg}^{1-\mathrm{m}}\mathrm{m}^{3-3\mathrm{m}}\mathrm{s}^{-1}]\,.$$

Therefore,

$$\varepsilon_1 = 2.69 \cdot 10^{-5}\,, \quad \varepsilon_2 = 2.74 \cdot 10^{-1}\,, \quad \varepsilon_3 = 2.42 \cdot 10^{-5}\,, \quad \varepsilon_4 = 5.22 \cdot 10^{-2}\,.$$

Specifically, we consider the initial-boundary value problem for model (6.5.11) linked to the following initial and boundary conditions:

$$u(0,x) = 0\,, \qquad u(t,0) = u(t,1) = 0\,. \tag{6.5.26}$$

Calculations related to Problem 6.5.1 are performed with $x_s = 0.5$ and $s(t) = t^2 \exp(-20t^2)$, the space discretization has been organized with $n_1 = n_2 = 151$ nodes on both the $\mathcal{D}_1$ and $\mathcal{D}_2$ domains, and the system of ordinary

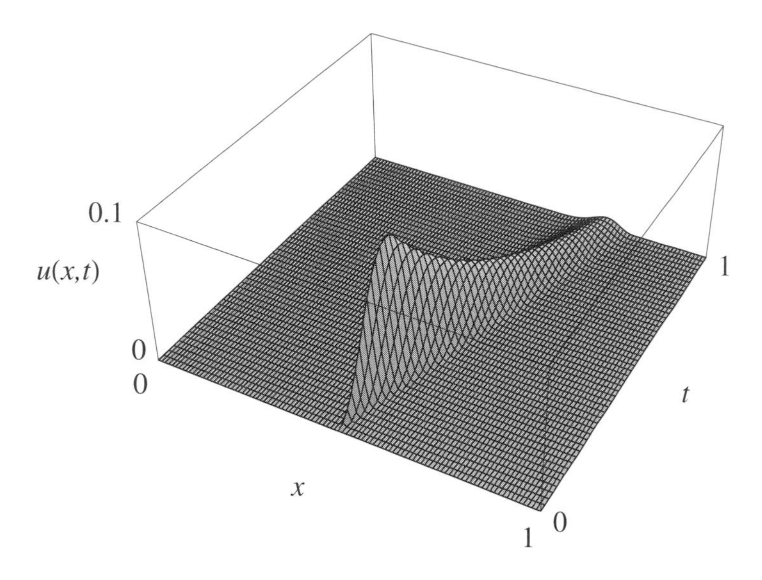

Figure 6.5.1.a - Temporal and spatial behavior of the concentration with a linear decay rate.

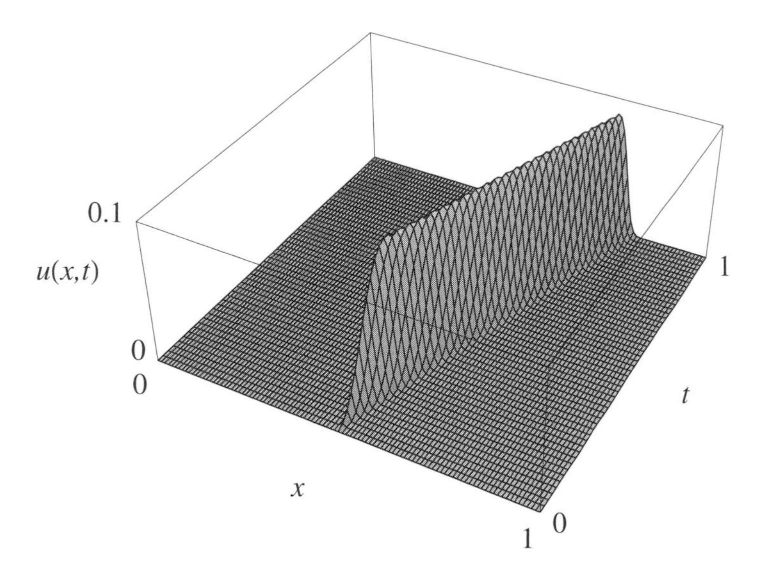

Figure 6.5.1.b - Temporal and spatial behavior of the concentration with a nonlinear decay rate.

differential equations has been integrated with a predictor-corrector method adopting an adaptative time step.

Figure 6.5.1.a shows the temporal and spatial behavior of the concentration u with a linear decay rate, that is $m = 1$ ($\mu = 3.24$), while Figure 6.5.1.b shows the temporal and spatial behavior of the concentration u with a nonlinear decay rate $m = 1.25$ ($\mu = 0.68$). Both figures illustrate that the algorithm is stable both in the peak zone of the solution and in the zone where the values of the variable u are very close to zero (typically near the boundary). The method turns out to be very efficient in terms of computational time: few minutes are sufficient to solve the ordinary differential nonlinear system of $n_1 + n_2 - 2$ equations with a common computer.

Problem 6.5.2 is analyzed with the choice of $x_s = 0.55$ while a measure downstream of the source position in $x^* = 0.565$ is chosen to be equal to $5t^2 \exp(-20t^2)$, i.e.,

$$u(x^*, t) = u(0.565, t) = 5t^2 \exp(-20t^2) \qquad \forall t \geq 0.$$

The space discretization is organized with $n_1 = 151$, $n_2 = 61$, and $n_3 = 151$ nodes on domains $\mathcal{D}_1$, $\mathcal{D}_2$, and $\mathcal{D}_3$, respectively, while the system of ordinary differential equations was integrated with a predictor-corrector method adopting an adaptative time step. In this case it is necessary to solve three different ordinary nonlinear systems of n_1, n_2, and n_3 equations, respectively.

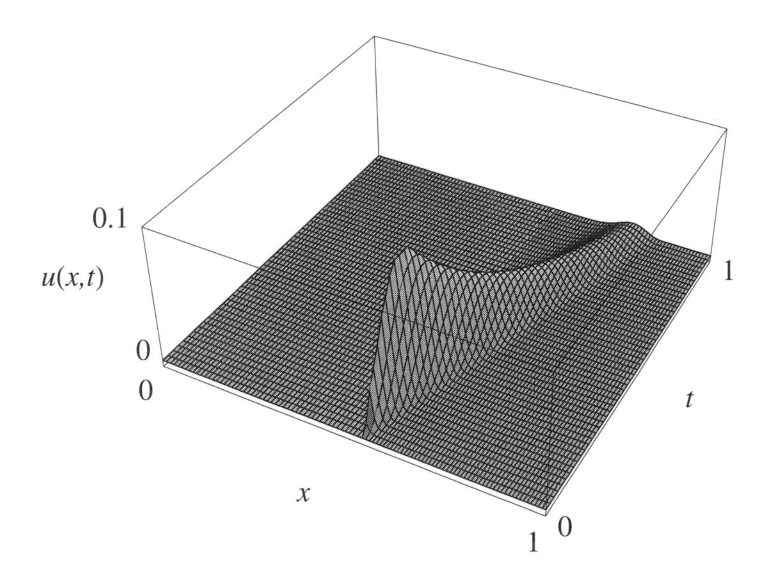

Figure 6.5.2.a - Evolution of the concentration.

Figure 6.5.2.a shows the behavior of the concentration for a linear decay term ($m = 1$ and $\mu = 3.24$) while Figure 6.5.2.b shows the reconstructed

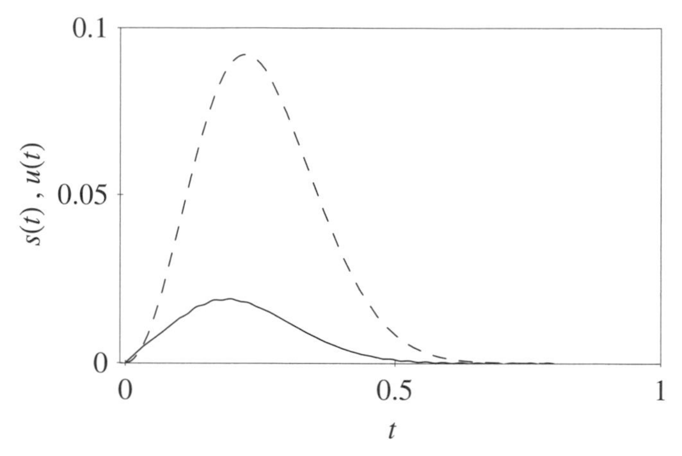

Figure 6.5.2.b - Reconstruction of the source term.

source (continuous line) compared to the measure in x^* (dashed line). In the same way, Figures 6.5.3.a and 6.5.3.b refer to a problem with a nonlinear decay term ($m = 2$ and $\mu = 0.00648$).

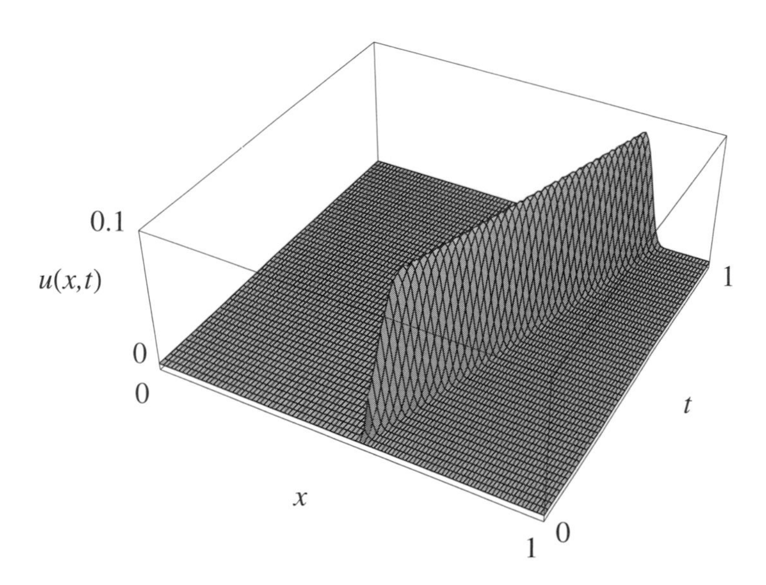

Figure 6.5.3.a - Evolution of the concentration.

Analogous results can be obtained for Problem 6.5.3 when the source is localized downstream of the point of the measure (i.e., in x_*). However, such results are less relevant from a physical point of view. Actually, because of the convective term, it would be necessary to localize the point of measure

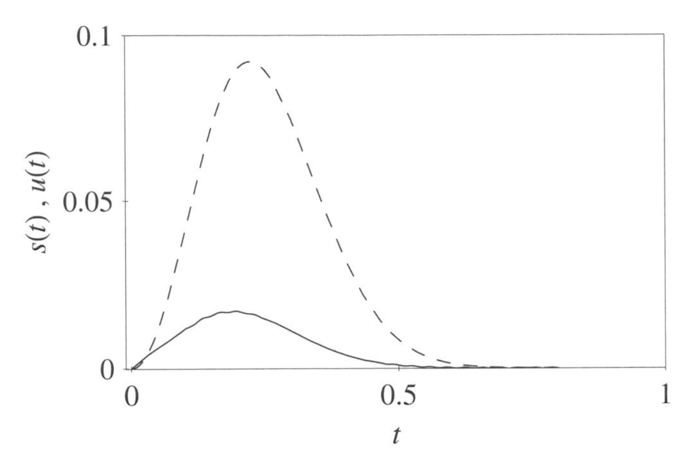

Figure 6.5.3.b - Reconstruction of the source term.

x_* very close to the source x_s corresponding to Problem 6.5.2 in order to show a significant concentration $u_*(t)$ (obviously this physical limitation disappears in a pure diffusive model).

6.6 Collocation Methods and Integro-Differential Equations

Generalized collocation methods can be developed to solve problems with integro-differential equations. This generalization is referred to a specific class of models, rather than to an abstract class of equations. We will deal with direct problems here; inverse problems, such as those seen above, can be approached by a technical development of the methods reported in the first part of this chapter.

Let us consider the mathematical model reported in Chapter 1 in Example 1.2.6, which is written, with reference to one population only, as follows:

$$\frac{\partial f}{\partial t}(t,u) = \int_0^1 \int_0^1 \eta(v,w)\psi(v,w;u) f(t,v) f(t,w)\, dv\, dw$$

$$-f(t,u) \int_0^1 \eta(u,w) f(t,w)\, dw\,, \tag{6.6.1}$$

with $u, v, w \in [0, 1]$, and where

$$\int_0^1 \psi(v, w, u)\, du = 1\,, \qquad \forall\, v, w \in [0,1] \times [0,1]\,. \tag{6.6.2}$$

The model is such that u has the physical meaning of a state variable at the microscopic scale, for an individual of a large population of anonymous interacting subjects; $\eta(v, w)$ is the encounter rate between individuals with state v and w, respectively, and $\psi(v, w; u)$ is the probability density that, after the encounter between individuals with state v and w, the individual with state v ends up in w. The dependent variable $f(t, u)$ is the probability density function over the microscopic state u.

The initial value problem is stated as follows:

$$\begin{cases} \dfrac{\partial f}{\partial t}(t, u) = J[f](t, u)\,, \\ f(0, u) = f_0(u)\,, \end{cases} \tag{6.6.3}$$

where $J[f]$ represents the right-hand side term of Eq. (6.6.1), and

$$\int_0^1 f_0(u)\, du = 1\,.$$

The existence and uniqueness of solutions to this problem have been proved in a paper by Arlotti and Bellomo (1995). The existence results are given in L^1 spaces for general initial conditions and under suitable assumptions on the terms η and ψ. Moreover, it is proved that the solution $f(t, \cdot)$ satisfies the following condition:

$$\int_0^1 f(t, u)\, du = 1\,, \quad \forall\, t \geq 0\,. \tag{6.6.4}$$

The application of the generalized collocation method will now be shown with reference to problem (6.6.3). Subsequently, some technical developments are dealt with. Consider first the collocation for the u-variable

$$I_u = \{u_1, \ldots, u_i, \ldots, u_m\}\,, \tag{6.6.5}$$

where $u_1 = 0$ and $u_m = 1$. Then, consider the approximation of f by interpolation,

$$f(t, u) \cong f^m(t, u) = \sum_{i=1}^{m} L_i(u) f_i(t). \tag{6.6.6}$$

Similarly, consider the probability density ψ which is interpolated with respect to the u-variable:

$$\psi(v, w; u) \cong \psi^m(v, w; u) = \sum_{i=1}^{m} L_i(u)\psi_i(v, w), \tag{6.6.7}$$

where $\psi_i(v, w) = \psi(v, w; u_i)$ while L_i denotes the ith-order fundamental interpolating functions, which can be Lagrange polynomials or sinc functions. It is useful to introduce different collocation sets for the v and the w-variables:

$$I_v = \{v_1, \ldots, v_p, \ldots, v_m\} \quad \text{and} \quad I_w = \{w_1, \ldots, w_q, \ldots, w_m\},$$

with $v_1 = w_1 = 0$ and $v_m = w_m = 1$; we assume here, for simplicity, that m is the common number of collocation points for all the u, v, and w-variables.

Moreover, let us introduce the following notation:

$$\eta_{pq} = \eta(v_p, w_q), \qquad \eta_{iq} = \eta(u_i, w_q), \tag{6.6.8}$$

and

$$\psi_{ipq} = \psi_i(v_p, w_q), \tag{6.6.9}$$

where $i, p, q = 1, \ldots, m$.

Substituting the above expressions in Eq. (6.6.1) yields:

$$\frac{df_i}{dt}(t) = H_i[f](t) - f_i(t)K_i[f](t), \tag{6.6.10}$$

which defines the time evolution of the values of the variable $f(t, \cdot)$ in the nodal points u_i.

The terms H_i and K_i define the approximation of the integral terms as follows:

$$H_i[f] = \sum_{p=1}^{m} \sum_{q=1}^{m} c_p\, c_q\, \eta_{pq}\, \psi_{ipq} f_p\, f_q, \tag{6.6.11}$$

and

$$K_i[f] = f_i \sum_{q=1}^{m} c_q\, \eta_{iq}\, f_q, \tag{6.6.12}$$

where $[f]$ denotes the set of the functions $\{f_i\}_{i=1}^{m}$, corresponding to the values of f in the discretization points, and

$$c_p = \int_0^1 L_p(v)\, dv, \qquad c_q = \int_0^1 L_q(w)\, dw. \tag{6.6.13}$$

The method leads to a system of ordinary differential equations which can be solved using the standard techniques we have seen in previous chapters. The difference is that it is not necessary to compute coefficients corresponding to space derivatives, while the coefficients (6.6.13) have to be computed in the early strings of the scientific program once the interpolating functions have been selected.

The above reasoning can be straightforwardly generalized to other kinds of models. For instance, models with a forcing term can be considered:

$$\left(\frac{\partial}{\partial t} + \sigma(t)\,\frac{\partial}{\partial u}\right) f(t,u)$$

$$= \int_0^1 \int_0^1 \eta(v,w)\psi(v,w;u) f(t,v) f(t,w)\, dv\, dw$$

$$- f(t,u) \int_D \eta(u,w) f(t,w)\, dw\,, \tag{6.6.14}$$

where $\sigma(t)$ is a known function of time.

In this case, the corresponding discrete system of ordinary differential equations can be written as follows:

$$\frac{df_i}{dt}(t) = -\sigma(t) \sum_{j=1}^{m} a_{ij} f_j + H_i[f](t) - f_i(t) K_i[f](t)\,, \tag{6.6.15}$$

where the expression of the coefficient $a_{ij} = a_{ij}^{(1)}$ $(i,j = 1,\ldots,m)$ is reported in Chapter 2, Section 2.2.

Moreover, boundary conditions have to be imposed at $u = 0$ or $u = 1$ using the same techniques we have seen for partial differential equations in the previous chapters.

Analogous reasoning applies to models with diffusion, with reference to Example 1.2.7, Chapter 1:

$$\left(\frac{\partial}{\partial t} + \varepsilon\,\frac{\partial^2}{\partial x^2}\right) f(t,x,u)$$

$$= \int_0^1 \int_0^1 \eta(v,w)\psi(v,w;u) f(t,x,v) f(t,x,w)\, dv\, dw$$

$$- f(t,x,u) \int_D \eta(u,w) f(t,x,w)\, dw\,. \tag{6.6.16}$$

where the dependent variable

$$f = f(t, x, u) \, : \, [0,1] \times [0,1] \times [0,1] \to \mathbb{R}_+ \, ,$$

stands for the probability density function of particles having position $x \in [0,1]$ and biological state $u \in [0,1]$ at time t.

In this case, the coefficients corresponding to the space derivatives have to be computed after discretization of the state variable u. Moreover, boundary conditions have to be imposed again using the same techniques introduced in the preceding chapters for partial differential equations.

Precisely, for any $(t, x) \in [0,1] \times [0,1]$, using the set

$$I_u = \{u_1, \ldots, u_i, \ldots, u_m\}$$

of collocation points for the u-variable, one has, as in Eq. (6.6.10),

$$\frac{df_i}{dt}(t, x) = H_i[f](t, x) - f_i(t, x) K_i[f](t, x) \qquad \forall i = 1, \ldots, m \, , \tag{6.6.17}$$

which defines the time evolution of the values of the variable $f(t, x, \cdot)$ in the nodal points u_i. The terms H_i and K_i define the approximation of the integral terms as in Eqs. (6.6.11) and (6.6.12). Now, introducing the set of collocation points for the x-variable

$$I_x = \{x_1, \ldots, x_n\} \, , \tag{6.6.18}$$

where $x = 0$ and $x_n = 1$, one approximates, for any $i = 1, \ldots, m$ and any $t \in [0,1]$, the function $f_i(t, x)$ as

$$f_i(t, x) \cong f_i^n(t, x) = \sum_{j=1}^{n} L_j(x) f_{ij}(t) \, , \tag{6.6.19}$$

where $f_{ij}(t)$ is the value of $f_i(t, x)$ in the nodal point x_j, namely

$$f_{ij}(t) = f(t, x_j, u_i) \, ,$$

for any $i = 1, \ldots, m$, $j = 1, \ldots, n$, and any $t \in [0,1]$.

With this method, the function f is interpreted as follows:

$$f(t, x, u) \cong f^{nm}(t, x, u) = \sum_{h=1}^{n} \sum_{i=1}^{m} L_h(x) L_i(u) f_{ih}(t) . \tag{6.6.20}$$

The corresponding discrete system is slightly different:

$$\frac{df_{ih}}{dt}(t) = -\varepsilon \sum_{k=1}^{n} a_{kh} f_{ik}(t) + H_{ih}[f](t) - f_{ih}(t) K_{ih}[f](t)\,, \qquad (6.6.21)$$

where the second-order derivative coefficients $a_{kh}^{(2)}$ are reported in Section 2.2 of Chapter 2. The expressions of the corresponding terms H_{ih} and K_{ih} follow from technical calculations left to the initiative of the reader.

The solution to problem (6.6.3) can be numerically tested, for instance, using condition (6.6.4). To do so, one has to compute the distance

$$\varepsilon_m(t) = \left| 1 - \frac{1}{m} \sum_{i=1}^{m} f_i(t) \right|, \qquad (6.6.22)$$

which provides an indication of the estimate of the computational error.

An additional test is obtained by comparing the computational solution to the analytic solution $f(t,u) = [f_0(u) - 1]e^{-t} - 1$ that corresponds to $\eta = 1$ and $\psi(v,w;u) = 1$, uniformly over the variables v and u.

Note that it is not difficult to extend the method to the unbounded domain D of the variable u, e.g., the positive real line $u \in D = [0,\infty)$, or the whole real line $u \in D = (-\infty,\infty)$. In this case it is natural to use sinc interpolation:

$$f(t,u) \cong f^m(t,u) = \sum_{i=-m}^{m} S_i(u) f_i(t)\,, \qquad (6.6.23)$$

while all related calculations are only technical generalizations of those developed for bounded domains.

6.7 Collocations and Orthogonal Approximation

The development of generalized collocation methods is based, as we have seen, on the use of fundamental interpolating functions, typically Lagrange polynomials or sinc functions. Mixed-type interpolations can be used for problems in two space dimensions. We have also discussed how one can improve accuracy by linking Lagrange-type interpolations to the proper collocation; the advantages and weaknesses have been discussed in various chapters of the book. These interpolations allow an immediate calculation of the space derivative in the collocation points. On the other hand, the interpolation error may not be monotone increasing with increasing number of nodes, while integration over time of the system of ordinary differential equations may propagate the error.

In general, it is useful to look for improved interpolation techniques, for instance, by using *spectral approximation* rather than interpolation techniques. In this case, the representation of a function $u = u(t, x)$ is given by an expansion of the type

$$u(t, x) = \sum_{i=-\infty}^{\infty} c_i(t) h_i(x) , \tag{6.7.1}$$

where the functions $h_i = h_i(x)$ belong to a complete space of orthogonal functions defined in a suitable Hilbert space with weighted inner product

$$\langle f \, , \, g \rangle_w(t) = \int f(t, x) g(t, x) w(x) \, dx , \tag{6.7.2}$$

where $w = w(x)$ denotes the weight function.

Therefore, it is not difficult to check that the coefficients c_i are given by

$$c_i(t) = \langle u(t, \cdot) \, , \, \psi_i(\cdot) \rangle_w , \tag{6.7.3}$$

where the calculations of the integrals may require numerical quadratures.

Various examples of orthogonal polynomials can be extracted from the specialized literature. We describe a few here.

Example 6.7.1
Legendre Polynomials

Legendre polynomials are characterized as follows:

- Range: $[-1, 1]$.
- Polynomials:

$$L_0(x) = 1 , \quad L_1(x) = x ,$$

and

$$L_m(x) = \frac{2m - 1}{m} x L_{m-1}(x) - \frac{m - 1}{m} L_{m-2}(x) . \tag{6.7.4}$$

- Weight:

$$w(x) = \frac{2}{2m + 1} . \tag{6.7.5}$$

□

Example 6.7.2
Laguerre Polynomials

Laguerre polynomials are characterized as follows:

- Range: $[0, \infty)$.
- Polynomials:

$$L_0^a(x) = 1\,, \quad L_1^a(x) = a + 1 - x\,, \quad a > -1\,, \tag{6.7.6}$$

and

$$L_m^a(x) = \frac{1}{m}(2m - 1 + a - x)L_{m-1}^a(x) - \frac{1}{m}(m - 1 + a)L_{m-2}^a(x)\,. \tag{6.7.7}$$

- Weight:

$$w(x) = \frac{m!}{\Gamma(a + m + 1)} x^a e^{-x}\,. \tag{6.7.8}$$

□

Example 6.7.3
Hermite Polynomials

Hermite polynomials are characterized as follows:

- Range: $(-\infty, \infty)$.
- Polynomials:

$$H_0(x) = 1\,, \quad H_1(x) = 2x\,,$$

and

$$H_{m+1}(x) = 2xH_{m-1}(x) - 2mH_m(x)\,. \tag{6.7.9}$$

- Weight:

$$w(x) = e^{-x^2}\,. \tag{6.7.10}$$

□

In addition, sinc functions, described in Chapter 2, can be used in the following range: $(-\infty, \infty)$, with weight $w(x) = 1$.

The function $u = u(t, x)$ can be ***approximated*** by the truncated expansion u^n corresponding to some space collocation $I_x = \{x_1, \ldots, x_m\}$

$$u(t, x) \cong u^n(t, x) = \sum_{i=-n}^{n} c_i(t)\psi_i(x)\,, \tag{6.7.11}$$

where the coefficients $c_i = c_i(t)$ can be computed numerically, exploiting the values of u in the collocation points

$$c_i^n(t) \cong \sum_{j=1}^{m} W_j u_j^n(t)\psi_i(x_j), \tag{6.7.12}$$

where W_j denotes the weight functions in the quadrature formula.

Let us now consider the solution to the following class of equations:

$$\frac{\partial u}{\partial t} = \eta(t,x)\,\frac{\partial u}{\partial x} + \mu(t,x)\,\frac{\partial^2 u}{\partial x^2} + \varepsilon\, f\left(t, x, u, \frac{\partial u}{\partial x}\right), \tag{6.7.13}$$

where η, μ, and f are given smooth functions of their arguments.

Substituting (6.7.11) into (6.7.13) yields

$$\frac{\partial}{\partial t}\left(\sum_{i=-n}^{n} c_i^n(t)\psi_i(x)\right) = \eta(t,x)\frac{\partial}{\partial x}\left(\sum_{i=-n}^{n} c_i^n(t)\psi_i(x)\right)$$

$$+\,\mu(t,x)\frac{\partial^2}{\partial x^2}\left(\sum_{i=-n}^{n} c_i^n(t)\psi_i(x)\right)$$

$$+\,\varepsilon f\left(t, x, \sum_{i=-n}^{n} c_i^n(t)\psi_i(x), \frac{\partial}{\partial x}\left(\sum_{i=-n}^{n} c_i^n(t)\psi_i(x)\right)\right). \tag{6.7.14}$$

The above equation, after evaluation of the inner product, can be formally written as follows:

$$\frac{dc_i^n}{dt} = \eta(t,x)I_{1i}(\mathbf{c},t) + \eta(t,x)I_{2i}(\mathbf{c},t) + \varepsilon I_{3i}(\mathbf{c},t), \tag{6.7.15}$$

where $\mathbf{c} = \{c_i\}_{i=1,\dots,n}$, and

$$I_{1i} = \sum_{i=-n}^{n}\left\langle \eta c_i^n \frac{d\psi_i}{dx}, \psi_i\right\rangle_w, \tag{6.7.16}$$

$$I_{2i} = \left\langle \mu c_i^n \frac{d^2\psi_i}{dx^2}, \psi_i\right\rangle_w, \tag{6.7.17}$$

and

$$I_{3i} = \left\langle f\left(t, x, \sum_{i=-n}^{n} c_i^n\psi_i, \sum_{i=-n}^{n} c_i^n \frac{d\psi_i}{dx}\right), \psi_i\right\rangle_w, \tag{6.7.18}$$

where the calculation of the above integral terms (6.7.16)–(6.8.18) requires the computation of the quadratures corresponding to the space collocation.

Equation (6.7.13) has to be implemented with initial conditions:

$$u_0(x) = u(0,x) \cong u^n(0,x) = \sum_{i=-n}^{n} c_{i0}\psi_i(x)\,, \tag{6.7.19}$$

where $c_{i0} = c_i(0)$. Therefore,

$$c_{i0}^n(t) \cong \sum_{j=1}^{m} W_j u_0(x)\,\psi_i(x_j)\,. \tag{6.7.20}$$

Let us consider the case of linear Dirichlet conditions. Imposing compatibility conditions yields

$$\alpha(t) = u(t,0) \cong \sum_{i=-n}^{n} c_i(t)\psi_i(0)\,, \tag{6.7.21}$$

and

$$\beta(t) = u(t,1) \cong \sum_{i=-n}^{n} c_i(t)\psi_i(1)\,, \tag{6.7.22}$$

while imposing Neumann boundary conditions yields

$$\gamma(t) = \frac{\partial u}{dx}(t,0) \cong \sum_{i=-n}^{n} c_i(t)\frac{\partial \psi_i}{dx}(0)\,, \tag{6.7.23}$$

and

$$\delta(t) = \frac{\partial u}{dx}(t,1) \cong \sum_{i=-n}^{n} c_i(t)\frac{\partial \psi_i}{dx}(1)\,. \tag{6.7.24}$$

The above relations represent a constraint for the coefficients c_i which can be implemented to the first and last equations of (6.7.15) similarly to the case of generalized collocation methods.

Additional analysis, as already mentioned in Chapter 2, is needed to use wavelet approximation methods, as documented in Daubechies (1992), Meyer and Ryan (1993), and more recently Cattani and Rushchitsky (2007), which provides technical aspects on the application of the method.

The above application to a specific class of equations has been selected as a simple test to show the implementation of the method. Generalizations to other types of equations are technical. The reader can develop suitable calculations and compare the preceding solution technique with the ones we have seen in the preceding chapters.

6.8 Critical Analysis

This chapter has shown how generalized collocation methods can be developed to deal with a large variety of nonlinear problems, focusing on to ill-posed problems and to the solution of integro-differential equations.

Applications have been referred to problems with nonlinear boundary conditions and problems with missing boundary conditions which are substituted by additional information on the solution of the problem. The technique has been further developed towards the solution of problems with a ***localized source term***: the solution method has shown how it is possible to identify the source term based upon measurements on the solution of the problem.

The method has been tested to deal with a problem concerning the model of diffusion of a pollutant described in Chapter 1. This application has clearly indicated the validity of the method to deal with a large variety of nonlinearities. Indeed, this chapter confirms both the validity and the limitations of the method. The method is shown to be efficient in dealing with nonlinear models with possibly nonlinear and time-dependent boundary conditions.

The above-mentioned efficiency also appears to be robust in the case of problems which need domain decomposition methods with matching of the solution at the boundary of contiguous domains. The application of Section 6.6 has shown that the above remarks can be applied to a large variety of problems.

On the other hand, the technical difficulties and the computational complexity we have faced in Chapter 5 with reference to problems in two space dimensions appear to be further emphasized for the complex problems in this chapter. Even the generalization of the method to deal with problems in two space dimensions is not straightforward and can effectively be performed only for several simple geometries. The above reasoning confirms once more the critical analysis already stated in the Preface of this book.

Finite difference methods require discretization of both time and space variables. Space derivatives are computed using the discretization points which are, in the grid, close to the node where the dependent variable is computed. The resulting computational scheme is algebraic rather than differential. It can be solved in the linear case by simple techniques with computational complexity generally lower than the one of collocation methods.

On the other hand, nonlinear problems need additional technical treatments such as local linearization or solution of nonlinear algebraic systems.

Technical methods to compute space derivatives heavily depend on the mathematical structure of the equation, say elliptic, parabolic, or hyperbolic. Several schemes are reported in Chapter 3 of the book by Bellomo and Preziosi (1996) with reference, as already mentioned, to the structure of the equation.

On the other hand, if the time variable is left continuous and space derivatives are computed as outlined above, the approach is that of the ***method of lines***, which is similar to the one developed in this book and is a relatively less general way of computing space derivatives.

This final chapter has been devoted to a variety of miscellaneous problems selected by the authors to show how the methods can be properly generalized towards the computational solution of several problems which appear technically different from those we have analyzed in the preceding chapters. Possibly additional generalizations and applications can be proposed, also considering that the preliminary analysis of this chapter has shown an effective flexibility of the method. The interested reader can exploit the hints given here towards further generalizations.

The contents of this chapter essentially confirm the critical analysis already proposed in the previous chapters, and already anticipated in the Preface. Indeed, the method is shown to be flexible enough to deal with a broad variety of problems; moreover, nonlinearities are technically controlled. Its application is easy and immediate.

An additional advantage of the method refers to the implementation of the boundary conditions which can be easily enforced even for time-dependent conditions. Moreover, it has been shown that the method operates efficiently even for nonlinear boundary conditions.

On the other hand, the method efficiently operates only in the case of two or three, at most, dependent variables, for instance time and one or two space variables.

The computational complexity generated by problems in several dependent variables can be simplified by domain decomposition methods to reduce the number of ordinary differential equations induced by discretization and interpolation.

The accuracy of computations has been discussed, in the preceding chapters, with reference to specific classes of problems. Also in the cases in this chapter, the analysis is left to computational experiments rather than to a direct theoretical convergence analysis.

The various topics in this chapter are concisely treated, leaving it to the reader, who may be interested in specific problems, to develop and apply them to the analysis of specific problems. In the perspective of opening a dialogue with applied scientists, we encourage the development of scientific programs which may enlarge the variety offered in the Appendix.

6.9 Problems

PROBLEM 6.1
Show how the solution method for problems with nonlinear boundary conditions simplifies in the case of Robin boundary conditions.

PROBLEM 6.2
Apply the solution method proposed in Section 6.3 to the traffic flow model.

PROBLEM 6.3
Develop a scientific program (with ***Mathematica***®) to deal with Problem 6.2 on the basis of the programming methods proposed in the previous chapters.

PROBLEM 6.4
Show how the solution method proposed in Section 6.4 can be technically generalized for a transport diffusion model.

PROBLEM 6.5
Show how the solution method proposed in Section 6.4 can be technically generalized for the traffic flow model.

PROBLEM 6.6
Develop a scientific program (with ***Mathematica***®) to deal with Problems 6.4, 6.5.

PROBLEM 6.7
Use the program mentioned in Problem 6.6 to analyze the sensitivity analysis to variable measurements.

PROBLEM 6.8
Develop a scientific program (with ***Mathematica***®) to deal with problem 6.5.1 of Section 6.5 and compare the computational solution with that shown in Figure 6.5.1.

PROBLEM 6.9
Develop a scientific program to deal with Problem 6.5.2 of Section 6.5 and compare the computational solution with that shown in Figure 6.5.2.

PROBLEM 6.10
Develop a scientific program to deal with Problem 6.5.3 of Section 6.5 and compare the computational solution with that shown in Figure 6.5.3.

Appendix
Scientific Programs

A.1 Introduction

This appendix provides a technical introduction to scientific programming based on ***Mathematica***®, which solves the examples and problems proposed throughout the book. Some of the programs could be used to potentially solve problems of much greater complexity than those identified by the specific problems or examples. The aim is also to direct the reader toward a natural use of ***Mathematica*** for the solution of problems in mechanical and applied sciences.

Before describing the programs, some technical aspects are given regarding their formal organization. Subsequently, the inner structure of the commands related to a specific program is described in detail. It is important to distinguish between two possible settings of the programs within the above framework. The first one, called ***Notebook***, presents the lines of the program on the screen for the user to view; the second one, called ***Package***, could be useful for very long programs. We assume that the reader has some knowledge, although not necessarily deep, of the main commands that are presented in the *Mathematica Book*. Therefore, the full list of these commands is not reported here; it is necessary to consult the handbook to complete the program.

The ***Notebook*** is structured as follows:

```
Program Name [input1_,...,inputn_]:=
Module [{ letters (symbols) },
Command 1;
⋮
Command m;
]
```

After the name of the program, the line in square brackets reports all the input data with the underscore. All the commands used in the program have to be contained between two square brackets preceded by the word ***Module***.

The letters between the curly brackets identify the meaning of the letters used in the framework of the program.

Type in the following line to run the program:

```
Program Name [input1,...,inputn]
```

where the underscore is no longer present.

Also note that the ***command lines end with a semicolon (;)***, while the command that recalls the program does not have any punctuation. The exact line that needs to be typed in is reported for the examples contained in this appendix.

The structure of the ***Package*** is as follows:

```
BeginPackage ["Program Name" ]
Program Name::usage:="description of the target of the
                                                program"
Begin ["`private`"]
Name of the program [input1_,...,inputn_]:=
Module [{ letters (symbols) },
Command 1;
⋮
Command m;
]
End[]
EndPackage[]
```

The Notebooks are ***saved on the hard disk of the personal computer***, as **name.nb**, while the Packages programs are ***saved*** as **name.m**.

Saving is achieved by using the **Save as Special** command, which is in the File menu. The above formal organization has to be followed carefully, otherwise the program does not run.

Two Notebooks are now explained in detail, **OneDLag.nb** used in Chapter 2 and **KdVIII.nb** used in Chapter 3.

The **OneDLag.nb** program shows the difference between the Chebychev and equally spaced collocations for a Lagrange interpolation.

The general command is

```
OneDLag[function, nodes]
```

in which the user must define the interpolated ***function*** and the number of ***nodes***. The particular case shown by the program is

```
OneDLag[Tanh[10x-5.]+0.2 Sin[π(10x-5.)],21]
```

The program writes

```
OneDLag[function_, Nodes_] := Module[{},
φ [x_] := function;
(*Definition of the equispaced and Chebychev collocation*)
hh = 1/(Nodes - 1);
x_i_ := (i - 1)*hh;
x1_i_ :=-(1/2)(Cos[(i - 1)*π/(Nodes - 1)] - 1);
(*Lagrange polynomial with Chebychev collocation*)
Lagr1[j_,x_] :=Product[If[p≠j,(x-x1_p)/(x1_j-x1_p),1],{p,1,Nodes}]
FuncLagr1[x_] :=Sum[((φ[x] /.x→ x1_k)*Lagr1[k, x]),{k,1,Nodes}]
(* Lagrange polynomial with equally-spaced collocation *)
Lagr2[j_,x_] :=Product[If[p≠j,(x-x_p)/(x_j-x_p),1],{p,1,Nodes}]
FuncLagr2[x_] :=Sum[((φ[x] /.x→ x_k)*Lagr2[k, x]),{k,1,Nodes}]
Plot[Evaluate[{φ[x], FuncLagr1[x], FuncLagr2[x]}], {x, 0, 1},
PlotRange → All, PlotStyle → {Dashing[{0.0}],
{Dashing[{0.001, 0.005}]}, {Dashing[{0.02, 0.008}]}}]
];
```

The sentences between the symbols $(* \cdots *)$ are comments to facilitate the understanding of the program.

hh is the equally spaced spatial step, while $x_{i\_}$ and $x1_{i\_}$ define two series of collocation points: the first one is equally spaced, the second one follows a Chebychev formulation.

The sentence Lagr1[$j\_, x\_$] defines the generic Lagrange polynomial related to the equally spaced collocation points x_i while Lagr2[$j\_, x\_$] defines the generic Lagrange polynomial related to the Chebychev collocation series $x1_i$.

Various options have been chosen relative to the ***Plot*** command. For instance, the ***Dashing*** options are chosen in order to draw the curves with different styles. These are shown together with the ***Show*** command.

The ***KdVII*** program performs the solution of a nonlinear initial value problem in an unbounded domain with special attention being paid to the solution of the third-order Korteweg–de Vries equation

$$\frac{\partial u}{\partial t} + u^m \frac{\partial u}{\partial x} + \mu \frac{\partial^3 u}{\partial x^3} = 0 \cdot$$

The general command is as follows:

```
KdVIII[IBData, Nodes, μ, m, Type]
```

where **IBData** is a vector containing the initial and boundary conditions, **Nodes** is the number of nodes, μ, **m** are two parameters, and **Type** is a choice between the sinc or Lagrange interpolation function.

The first part of the program is very similar to the previous one. The differences are in its central part, which is written as follows:

```
⋮
(*Derivative Matrices*)
If[Type ===Lagrange,FDM[j_, i_]:=Which[i == j,
Sum[If[k==i,0,1/(x1i-x1k)],{k,1,Nodes}],i!=j,
(Product[If[p==i,1,x1i-x1p],{p,1,Nodes}])/((x1i-x1j)*
Product[If[p==i,1,x1j-x1p],{p,1,Nodes}])];
FirstDer=Table[FDM[j, i], j, 1,Nodes, i, 1, Nodes];
SecondDer=Table[If[ij,2(FirstDer[[j,i]]*FirstDer[i,i]]-
FirstDer[[j, i]]/(x1i-x1j),0],{j,1,Nodes},{i,1,Nodes}];
Do[SecondDer[[i, i]]=-Sum[SecondDer[[k,i]],{i,1,Nodes}];
ThirdDer=Table[If[ij,3(FirstDer[[j,i]]*SecondDer[[i,i]]-
SecondDer[[j, i]]/(x1i-x1j),0],{j,1,Nodes},{i,1, Nodes}];
Do[ThirdDer[[i, i]]=-Sum[ThirdDer[[k,i]],{i,1,Nodes}];
If[Type===Sinc,
FDM[j_,i_]:=Which[i!=j,(-1)^(i-j)/(hh*(i-j)),i==j,0];
SDM[j_,i_]:=Which[i!=j,2*(-1)^(i-j+1)/
(hh^2*(i - j)^2),i==j,-π^2/(3*hh^2)];
TDM[j_,i_]:=Which[i!=j, (-1)^(i-j)/hh^3(6/(i-j)^3-π^2/(i-j)),i==j,0];
FirstDer=Table[FDM[j, i],{j,1,Nodes},{i,1,Nodes}];
SecondDer=Table[SDM[j, i],{j,1,Nodes},{i,1,Nodes}];
ThirdDer=Table[TDM[j, i],{j,1,Nodes},{i,1,Nodes}]];
(*Nodal Equations*)
Ci'[t]==((-(T1-T0)/(X1-X0))*(Sum[FirstDer[[k,i]]*Ck[t]),
{k,1,Nodes}]*Ci[t]^mreact+(-μ(T1-T0)/(X1-X0)^3)*
(Sum[ThirdDer[[k,i]]*Ck[t]),{k,1,Nodes}]))/.x→x1i;
EqsSys = Flatten[Table[{NdEq[i], Ci[0] == (φ[x] /. x→ x1i)},
{i, 2, Nodes - 1}]];
NdEqsSys = Join[EqsSys, {C1'[t] ==D[InitBoundData[[1]], t],
C1[0] ==(φ[x]/.x→x11),CNodes'[t] == D[InitBoundData[[3]],t],
CNodes[0] ==φ[x]/.x→x1Nodes)}];
UnKnown = Join[Table[Ci,{i, 1, Nodes}]];
NumSol=NDSolve[NdEqsSys,UnKnown,{t,0,1},MaxSteps→10^8]
                                                    //Flatten;
```

```
(*Solution*)
Which[Type === Sinc, TotSol[x_, t_] := Sum[NumSol[[k, 2]][t]*
Sinc[k, x],{k,1,Nodes}], Type === Lagrange,TotSol[x_, t_] :=
Sum[NumSol[[k, 2]][t]*Lagr[k, x],{k,1,Nodes}]]
⋮
```

FDM, ***SDM***, and ***TDM*** are the first, second, and third coefficients of the space derivative matrices, respectively. ***FirstDer***, ***SecondDer***, and ***ThirdDer*** collect these coefficients in three matrices. This operation is performed to increase the efficiency of the scientific program. ***NdEq*** writes the generic ordinary differential equation which defines the time evolution of the value of the variable ***C*** in nodal points. The equations are then linked, in ***NdEqsSys***, to the boundary conditions. ***Unknown*** is a table that contains a list of the unknown variables. Finally, ***NumSol*** performs the ***NDSolve*** command that solves the system of ordinary differential equations with given initial data by using the option of maximum number of time steps given by ***MaxSteps***.

The following sections provide a detailed description of the scientific programs used in this book. We show the scope of each program and indicate how to solve the examples (mathematical problems) reported in the various chapters of the book.

The reader will not find a detailed description of each specific command, but only a brief explanation of the general input commands, the input data, and the output results. In particular, all the variables in the general command line have to be substituted by numbers or symbolic expressions, either before whatever calls the Notebook, followed by a semicolon (;), or in the calling itself. The meaning of all the variables is then given in an easily consulted scheme. References are sometimes made to the sections that give the theoretical basis needed to understand the output of the program and to other applications of the program already discussed in the book. A background, although not necessarily deep, on the use of ***Mathematica*** is clearly necessary. Moreover, the reader is encouraged to put great effort into performing the changes in the programs necessary to solve the problems proposed in the book.

A.2 Lagrange Interpolation with Chebychev and Equally Spaced Collocations

The **OneDLag.nb** notebook has the purpose of showing the difference between the Chebychev and the equally spaced collocation in the case of Lagrange interpolation (see ***Example 2.4.1***). The general command is

```
OneDLag[function, nodes]
```

where

function: interpolating function

nodes: number of nodes.

The result is given in the form of a figure where the analytical expression of the function and its Chebychev and equally spaced collocations are compared. For example, in the case of the function

$$u(x) = \tanh\left[5(2x-1)\right] + \frac{1}{5}\sin\left[5\pi(2x-1)\right]$$

the list of commands is

```
{function,nodes}={Tanh[10x-5.]+0.2 Sin[π(10x-5.)],21};
OneDLag[function, nodes]
```

A.3 Lagrange and Sinc Interpolation in One Space Dimension

The **OneDLaSiInt.nb** Notebook illustrates a comparison between the Lagrange and sinc interpolations in the case of a function in one space variable. The Chebychev collocation is adopted for the Lagrange interpolation, while the program develops an equally spaced collocation for the sinc interpolation (see ***Example 2.4.2***). The general command is as follows:

```
OneDLaSiInt[function, nodes]
```

where

function: interpolating function

nodes: number of nodes.

The result is given in a series of figures, where the analytical form of the function and its sinc and Lagrange interpolations are compared. The program also compares the graphics of the first and second derivative functions using a direct derivation of the sinc and Lagrange polynomials.

The Notebook shows two examples in order to demonstrate the different capabilities of the sinc and Lagrange interpolations:

- *Gaussian-like form.* The function is

$$v(x) = \exp\left[-50(2x-1)^2\right]$$

and the list of commands is

```
{function,nodes}={Exp[-2(10x-5]^2],21};
OneDLaSiInt[function, nodes]
```

- *Wavy hyperbolic tangent.* The function is

$$v(x) = u(x) = \tanh\left[5(2x-1)\right] + \frac{1}{5}\sin\left[5\pi(2x-1)\right]$$

and the list of commands is as follows:

```
{function,nodes}={Tanh[10x-5.]+0.2 Sin[π(10x-5.)],21};
OneDLaSiInt[function, nodes]
```

A.4 Error Computation in One Space Dimension

This program (***OneDLSerr.nb***) performs the computation of the maximum and mean errors delivered by a sinc and a Lagrange interpolation with an equally spaced and a Chebychev collocation, respectively (see ***Example 2.4.3***). The points where the error is computed are taken equally spaced between 0 and 1 with a step denoted by *delta*.

The program creates a table listing the differences between the analytical value of the function and interpolation in these points. The maximum and the mean values of this table are then computed. The Notebook performs the calculation from a minimum number of nodes, $MinNodes$, to a maximum number of nodes, $MaxNodes$. The results are stored in two arrays: `ErrorMax` and `ErrorMean`. The general command is as follows:

```
OneDLaSiErr[function, nodes, delta]
```

where

function: interpolating function

nodes: number of nodes

delta: incremental step in the spatial direction.

The Notebook shows two examples in order to demonstrate the different capabilities of the sinc and Lagrange interpolations:

- *Gaussian-like form.* The function is

$$v(x) = \exp\left[-50(2x-1)^2\right]$$

and the list of commands is

```
{function,MinNodes,MaxNodes,delta} =
                              {Exp[-2(10x-5]^2],2,41,0.001};
For[i = MinNodes,i ≤ MaxNodes,
                    OneDLaSiErr[function, i, delta];i=i+1];
```

- *Wavy hyperbolic tangent.* The function is

$$v(x) = u(x) = \tanh[5(2x-1)] + \frac{1}{5}\sin[5\pi(2x-1)]$$

and the list of commands is

```
{function,MinNodes,MaxNodes,delta}=
                    {Tanh[10x-5.]+0.2 Sin[π(10x-5.)],2,41.0.001};
For[i = MinNodes,i ≤ MaxNodes,
                        OneDLaSiErr[function, i, delta];i=i+1];
```

A.5 Gaussian Function in Two Space Dimensions

The **TwoDLaSiInt.nb** Notebook compares Lagrange and sinc interpolations in the case of functions of two space variables. The collocation for the Lagrange interpolation is a Chebychev one while the program performs an equally spaced collocation for the sinc interpolation (see ***Example 2.4.4***). The program allows different numbers of nodes to be dealt with in the x or y directions. The general command is

```
TwoDLaSiInt[function, NodesX, NodesY]
```

where

function: interpolating function

NodesX: number of nodes in the x direction

NodesY: number of nodes in the y direction.

The result is visualized by two figures where the sinc and Lagrange interpolations are represented. The program also plots the graphics of the differences between the analytical function and the interpolated one.

The Notebook uses, as an example, the Gaussian-type function

$$w(x,y) = \exp\left[-25(2x-1)^2 - 25(2y-1)^2\right]$$

and the list of commands is

```
{function,NodesX,NodesY}={Exp[-25(2x-1]^2-25(2y-1]^2],15,15};
TwoDLaSiInt[function,NodesX,NodesY]
```

A.6 Error Computation in Two Space Dimensions

The ***TwoDLaSiErr.nb*** program computes the maximum and mean error given by a sinc and a Lagrange interpolation with an equally spaced and a Chebychev collocation, respectively (see ***Example 2.4.4***), in a two-dimensional space. The points, along the x-axis (respectively the y-axis) where the error is computed are taken between 0 and 1 with a step of *deltax* (respectively *deltay*).

The program creates a table listing the differences between the analytical value of the function and its interpolations and computes the maximum and the mean values of this table. The Notebook performs the calculation from a minimum number of nodes, $MinNodes$, to a maximum number of nodes, $MaxNodes$, taken to be equal in both directions. The results are stored in two arrays: `ErrorMax` and `ErrorMean`. The general command is

```
TwoDLaSiErr[function, nodes, deltax, deltay]
```

where

function: interpolating function

nodes: number of nodes in the x and y directions

deltax: incremental step in the x spatial direction

deltay: incremental step in the y spatial direction.

The Notebook shows, as an example, the error evaluation for the Gaussian function already used in the previous program

$$w(x,y) = \exp\left[-25(2x-1)^2 - 25(2y-1)^2\right] ,$$

and the list of commands is

```
{function,MinNodes,MaxNodes,deltax,deltay}=
                         {Exp[-25(2x-1]^2-25(2y-1]^2],15,15};
For[i = MinNodes,i ≤ MaxNodes,
           TwoDLaSiErr[function,i,i,deltax,deltay];i=i+1];
```

A.7 Solution of the Third-Order KdV Model

The **KdVIII.nb** Notebook performs the solution of a nonlinear initial value problem in unbounded domains with special attention devoted to the solution of the third-order Korteweg–de Vries equation, as stated in ***Section 3.3***. The model is

$$\frac{\partial u}{\partial t} + u^m \frac{\partial u}{\partial x} + \mu \frac{\partial^3 u}{\partial x^3} = 0 \cdot$$

The general command is given by

```
KdVIII[IBData, Nodes, μ, m, Type]
```

where

IBData: vector containing the initial and boundary conditions

Nodes: number of nodes

μ, **m:** two parameters

Type: choice between sinc or Lagrange interpolation functions.

The program performs two examples:

- *Solitary wave solution.* The initial condition is considered

$$u_0(x) = [A\,\mathrm{sech}^2(kx - x_0)]^{1/m}\,,$$

where

$$A = 2\mu k^2 \frac{(m+1)(m+2)}{m^2}\,.$$

The list of commands is as follows:

```
{T0,T1,X0,X1}={0,40,-30,30};
{Nodes,m,μ,k,x0}={51,1,1.,0.3,0};
{A,ω}={2(m+1)(m+2)(m^-2)μk^2,4μk^3m^-2};
InitCond=(A Sech[k((X1-X0)x+X0)-ω(T1-T0)t]^2-x0)^(1/m)//.t→ 0;
Bound0Cond=(A Sech[k((X1-X0)x+X0)-ω(T1-T0)t]^2-x0)^(1/m)//.x→ 0;
Bound1Cond=(A Sech[k((X1-X0)x+X0)-ω(T1-T0)t]^2-x0)^(1/m)//.x→ 1;
IBData={Bound0Cond,InitCond,Bound1Cond};
KdVIII[IBData, Nodes, μ, m, Type]
```

- *Two solitary wave solution.* We consider, in this case, the solution

$$u(t,x) = \sum_{i=1}^{2} \left[A_i\,\mathrm{sech}^2\left(k_i x - \omega_i t - x_i\right)\right]^{1/m}$$

where $A_i = 2(m+1)(m+2)m^{-2}\mu k_i^2$, $\omega_i = 4\mu k_i^3 m^{-2}$, and k_i, x_i are real numbers. The boundary conditions are taken to be identically equal to zero and the initial condition is

$$u_0(x) = \sum_{i=1}^{2} \left[A_i\,\mathrm{sech}^2\left(k_i x - x_i\right)\right]^{1/m}\,.$$

Now the list of commands is given by

```
{T0,T1,X0,X1}={0,360,-70,70};
```

```
{Nodes,m,μ,k1,k2,x1,x2}={81,2,1.,0.3,0.2,-2,3};
{A1,A2,ω1,ω2}={2(m+1)(m+2)(m^-2)μk1^2,
                        2(m+1)(m+2)(m^-2)μk2^2,4μk1^3m^-2,4μk2^3m^-2};
InitCond=(A1 Sech[k1((X1-X0)x+X0)-ω1(T1-T0)t-x1]^2)^(1/m)+
        (A2 Sech[k2((X1-X0)x+X0)-ω2(T1-T0)t-x2]^2)^(1/m)//.t→ 0;
Bound0Cond=0
Bound1Cond=0
IBData={Bound0Cond,InitCond,Bound1Cond};
KdVIII[IBData, Nodes, μ, m, Type]
```

The program also computes the invariants $I_1, \ldots, I_5$, as reported in ***Section 3.3***.

A.8 Solution of the Fifth-Order KdV Model

The **KdVV.nb** Notebook performs the solution of a nonlinear initial value problem in an unbounded domain and, specifically, the solution of the fifth-order Korteweg–de Vries equation, as in ***Section 3.4***. The model is as follows:

$$\frac{\partial u}{\partial t} + u\frac{\partial u}{\partial x} + \frac{\partial^3 u}{\partial x^3} - \frac{\partial^5 u}{\partial x^5} = 0\cdot$$

The general command is given by

```
KdVV[IBData, Nodes, Type]
```

where

IBData: vector containing the initial and boundary conditions

Nodes: number of nodes

Type: choice between sinc or Lagrange interpolation function

The program performs an example relative to the solitary wave solution; the initial condition is

$$u_0(x) = \frac{105}{169}\mathrm{sech}^4\left[\frac{x - x_0}{2\sqrt{13}}\right],$$

where x_0 is a free parameter. The boundary conditions are taken identically equal to zero. The list of commands is

```
{T0,T1,X0,X1}={0,100,-30,60};
{Nodes,x0}={61,0};
InitCond=105/169 Sech[1/(2√13)(((X1-X0)x+X0)-x0-
                              36/169((T1-T0)t+T0))]^4)//.t→ 0;
```

```
Bound0Cond=0;
Bound1Cond=0;
IBData={Bound0Cond,InitCond,Bound1Cond};
KdVV[IBData, Nodes, Type]
```

A.9 Traffic Flow Model (1 Variable) with Dirichlet Boundary Conditions

The **Traffic1.nb** Notebook develops the solution method described in ***Section 4.2*** for the initial-boundary value problem with Dirichlet boundary conditions for the traffic flow (***Section 4.3***):

$$\frac{\partial u}{\partial t} = (2u-1)\frac{\partial u}{\partial x} + \eta u^2(1-u)\frac{\partial^2 u}{\partial x^2} + \eta u(2-3u)\left(\frac{\partial u}{\partial x}\right)^2 .$$

The general command is given by

```
Traffic1[IBData, Nodes, η, Type]
```

where

IBData: the vector containing initial and boundary conditions

Nodes: number of nodes

η**:** parameter

Type: choice between sinc or Lagrange interpolation function

The program shows one example with Dirichlet boundary conditions. The list of commands is

```
{Nodes,η}={11,2.};
InitCond=0.2;
Bound0Cond=0.5-0.3e^-t;
Bound1Cond=0.2;
IBData={Bound0Cond,InitCond,Bound1Cond};
Traffic1[IBData, Nodes, η, Lagrange]
```

A.10 Traffic Flow Model (2 Variables) with Dirichlet Boundary Conditions

The **Traffic2.nb** Notebook solves a problem analogous to that dealt with by the previous program, namely an initial-boundary value problem with Dirichlet boundary conditions. However, the model now has two dependent variables (u, q). Still referring to ***Section 4.3***, the model under consideration is defined by

$$\begin{cases} \dfrac{\partial u}{\partial t} = -\dfrac{\partial q}{\partial x}\,, \\ \dfrac{\partial q}{\partial t} \left[(1-2u) + \eta u(2-3u)\dfrac{\partial q}{\partial x}\right]\dfrac{\partial q}{\partial x} + \eta u(1-u)\dfrac{\partial^2 q}{\partial x^2}\,. \end{cases}$$

The general command is given by

```
Traffic2[IBDataU,IBdataQ, Nodes, η, Type]
```

where

IBDataU: vector containing the initial and boundary conditions for the variable u

IBDataQ: vector containing the initial and boundary conditions for the variable q

Nodes: number of nodes

η**:** parameter

Type: choice between sinc or Lagrange interpolation function

Note that even when the model does not require boundary conditions for the u variable, the Notebook is able to include boundary conditions. In this case, the information is irrelevant (we have put two zeros), but it is very simple to change the Notebook in order to consider a problem with the proper boundary conditions.

The program shows the result for ***Problem 4.3.2***, and the list of commands, in this case, is given by

```
{Nodes,η}={11,1.};
{InitCondU,InitCondQ}={0.2,0.16};
{Bound0CondU,Bound0CondQ}={0,0.16};
{Bound1CondU,Bound1CondQ}={0,0.16+0.1 Sin[5 t]};
IBDataU={Bound0CondU,InitCondU,Bound1CondU};
IBDataQ={Bound0CondQ,InitCondQ,Bound1CondQ};
Traffic2[IBDataU, IBDataQ, Nodes, η, Lagrange]
```

A.11 Nonlinear Diffusion Model with Neumann Boundary Conditions

The **DiffNeu.nb** Notebook performs the numerical solution of a nonlinear diffusion model with Neumann boundary conditions. We consider the heat diffusion model, as in ***Section 4.4*** and ***Problem 4.4.1***:

$$\frac{\partial u}{\partial t} = u(1-u)\frac{\partial^2 u}{\partial x^2} + (1-2u)\left(\frac{\partial u}{\partial x}\right)^2 .$$

The general command is

```
DiffNeu[IBDataU, Nodes, Type]
```

where

IBData: vector containing the initial and boundary conditions

Nodes: number of nodes

Type: choice between sinc or Lagrange interpolation function.

As an example, the Notebook shows the case of natural Neumann boundary conditions and a biquadratic initial condition. The list of commands is as follows:

```
Nodes=21;
InitCond=4x^2(x-1)^2;
Bound0Cond=0;
Bound1Cond=0;
IBData={Bound0Cond,InitCond,Bound1Cond};
DiffNeu[IBData, Nodes, Lagrange]
```

A.12 Nonlinear Diffusion Model with Robin Boundary Conditions

The **DiffRob.nb** Notebook shows the numerical solution of a nonlinear diffusion model with Robin boundary conditions. The solution refers to the heat diffusion model, as in ***Section 4.4*** and ***Problem 4.4.2***:

$$\frac{\partial u}{\partial t} = u(1-u)\frac{\partial^2 u}{\partial x^2} + (1-2u)\left(\frac{\partial u}{\partial x}\right)^2 .$$

The general command is

```
DiffRob[IBDataU, Nodes, Type]
```

where

IBData: vector containing the initial and boundary conditions

Nodes: number of nodes

Type: choice between sinc or Lagrange interpolation function.

As an example, the Notebook illustrates the case of Robin boundary conditions with a linear combination between the Dirichlet and Neumann conditions (where all the constants are taken to be equal to 1) and a bi-quadratic initial condition. The list of commands is as follows:

```
Nodes=21;
InitCond=4x^2(x-1)^2;
{const1,const2,const3,const4}={1,1,1,1};
Bound0Cond=0;
Bound1Cond=0;
IBData={Bound0Cond,InitCond,Bound1Cond};
DiffRob[IBData, Nodes, Lagrange]
```

A.13 Problems with Known Analytic Solutions

The **ConvLaSi.nb** Notebook displays the numerical simulation for a problem with a known analytic solution. It deals specifically with a linear convection diffusion model with a source term (see ***Problem 4.5.1***). The model is

$$\frac{\partial u}{\partial t} + \frac{\partial u}{\partial x} = \frac{\partial^2 u}{\partial x^2} + s(t, x),$$

where $s(t, x)$ is the source term. The general command is

```
ConvLaSi[IBDataU, Nodes, Type]
```

where

IBData: vector containing the initial and boundary conditions

Nodes: number of nodes

Type: choice between sinc or Lagrange interpolation function.

As an example, the Notebook illustrates the case of a linear convection diffusion model with a source given by

$$s(t, x) = \exp[(1-x)^2 - t][-5 + 2x - 4(1-x)^2] .$$

The list of commands is as follows:

```
Nodes=21;
InitCond=Exp[(1-x)^2];
Bound0Cond=Exp[(1-t)];
Bound1Cond=Exp[(-t)];
IBData={Bound0Cond,InitCond,Bound1Cond};
ConvLaSi[IBData, Nodes, Lagrange]
ConvLaSi[IBData, Nodes, Sinc]
```

A.14 Initial-Boundary Value Problem on a Slab

The **SlabTwoDim.nb** Notebook deals with the simulation of initial-boundary value problems in two space dimensions, focused on problems in a slab geometry, as investigated in ***Section 5.4***. The model is defined by

$$\frac{\partial u}{\partial t} = (1-2u)\left[\left(\frac{\partial u}{\partial x}\right)^2 + \left(\frac{\partial u}{\partial y}\right)^2\right] + u(1-u)\left[\frac{\partial^2 u}{\partial x^2} + \frac{\partial^2 u}{\partial y^2}\right].$$

The general command is

```
SlabTwoDim[IBDataX,IBDataY,NodesX,NodesY,TypeX,TypeY]
```

where

IBDataX: vector containing the initial and boundary conditions for the x-variable

IBDataY: vector containing the initial and boundary conditions for the y-variable

NodesX: number of nodes in the x direction

NodesY: number of nodes in the y direction

TypeX: choice between sinc or Lagrange interpolation function in x direction

TypeY: choice between sinc or Lagrange interpolation function in y direction

The example reported in the Notebook shows the solution with sinc interpolation in the x direction and Lagrange interpolation in the y direction. The list of commands is

```
{NodesX,NodesY}={31,31};
{CoeffA,CoeffB,LL}={.1,1,40};
conciniz=Exp[-CoeffA*(LL(x-0.5))^2];
concBoundX0=0;
concBoundX1=0;
concBoundY0=Exp[-CoeffA*(LL(x - 0.5)^2]Exp[-CoeffB*t];
concBoundY1=Exp[-CoeffA*(LL(x - 0.5)^2]Exp[-CoeffB*t];
IBDataX={concBoundX0,conciniz,concBoundX1};
IBDataY={concBoundY0,conciniz,concBoundY1};
{TypeX,TypeY}={Sinc,Lagrange};
SlabTwoDim[IBDataX,IBDataY,NodesX,NodesY,TypeX,TypeY]
```

A.15 Heat Equation over a Square

The **HeatTwoDim.nb** Notebook performs the simulation for two-dimensional diffusion problems over a finite domain (see ***Section 5.5***). More specifically, the Notebook deals with the following initial-boundary value problem over a square:

$$\frac{\partial u}{\partial t} = \frac{1}{\pi^2}\left(\frac{\partial^2 u}{\partial x^2} + \frac{\partial^2 u}{\partial y^2}\right).$$

The general command is

```
HeatTwoDim[IBDataX,IBDataY,NodesX,NodesY,TypeX,TypeY]
```

where

IBDataX: vector containing the initial and boundary conditions for the x-variable

IBDataY: vector containing the initial and boundary conditions for the y-variable

NodesX: number of nodes in the x direction

NodesY: number of nodes in the y direction

TypeX: choice between sinc or Lagrange interpolation function in the x direction

TypeY: choice between sinc or Lagrange interpolation function in the y direction

The example reported in the Notebook shows the numerical solution of a problem with Dirichlet boundary conditions that admits an analytical solution. In this way the initial and boundary conditions are given by the analytical solution. The list of commands is

```
{NodesX,NodesY}={11,11};
Anal[x_,y_,t_]:=Exp[-t/2]Cos[π(x+y)/2]+Exp[-2t]sin[π(x-y)];
conciniz=Anal[x,y,0];
concBoundX0=Anal[0,y,t];
concBoundX1=Anal[1,y,t];
concBoundY0=Anal[x,0,t];
concBoundY1=Anal[x,1,t];
IBDataX={concBoundX0,conciniz,concBoundX1};
IBDataY={concBoundY0,conciniz,concBoundY1};
{TypeX,TypeY}={Lagrange,Lagrange};
SlabTwoDim[IBDataX,IBDataY,NodesX,NodesY,TypeX,TypeY]
```

A.16 Reaction-Diffusion Equations

The **EvolPatt.nb** Notebook deals with the simulation of an initial-boundary value problem in two space dimensions with Neumann boundary conditions, like the one investigated in ***Section 5.6***. The model is

$$\begin{cases} \dfrac{\partial u}{\partial t} = \kappa(a - u + u^2 v) + \varepsilon\left(\dfrac{\partial^2 u}{\partial x^2} + \dfrac{\partial^2 u}{\partial y^2}\right), \\ \dfrac{\partial v}{\partial t} = \kappa(b - u^2 v) + \left(\dfrac{\partial^2 v}{\partial x^2} + \dfrac{\partial^2 v}{\partial y^2}\right). \end{cases}$$

The general command is

```
EvolPatt[BDataUX,BDataVX,BDataUY,BDataVY,IData,
                                NodesX,NodesY,TypeX,TypeY]
```

where

BDataUX: vector containing boundary conditions for the u variable in the x direction

BDataVX: vector containing boundary conditions for the v variable in the x direction

BDataUY: vector containing boundary conditions for the u variable in the y direction

BDataVY: vector containing boundary conditions for the v variable in the y direction

IData: vector containing the initial condition

NodesX: number of nodes in the x direction

NodesY: number of nodes in the y direction

TypeX: choice between sinc or Lagrange interpolation function in the x direction

TypeY: choice between sinc or Lagrange interpolation function in the y direction.

The example reported in the Notebook shows the Schnakenberg reaction-diffusion model with natural Neumann boundary conditions. The list of commands is

```
{NodesX,NodesY}={11,11};
{D1,D2,κ,a,b,TT}={0.05,1,100,0.1305,0.7695,2};
concinizV=b/(a+b)^2;
concinizU=a+b+10^-3Exp[-100((x-1/3)^2+(y-1/2)^2)];
BDataUX={0, 0};
BDataUY={0, 0};
```

```
BDataVX={0, 0};
BDataVY={0, 0};
IData={concinizU, concinizV};
{TypeX,TypeY}={Lagrange,Lagrange};
EvolPatt[BDataUX,BDataVX,BDataUY,BDataVY,IData,
                                NodesX,NodesY,TypeX,TypeY]
```

A.17 Vehicular Traffic Model with Nonlinear Boundary Conditions

The **DiffNL.nb** Notebook performs the numerical solution of a nonlinear diffusion model with nonlinear boundary conditions. The model is the vehicular traffic model, like that in ***Section 6.3***, ***Example 6.3.1***,

$$\frac{\partial u}{\partial t} + (1-2u)\frac{\partial u}{\partial x} = \eta u^2(1-u)\frac{\partial^2 u}{\partial x^2} + \eta u(2-3u)\left(\frac{\partial u}{\partial x}\right)^2$$

and the general command is

```
DiffNL[IBDataU, Nodes, η, Type]
```

where

IBData: vector containing the initial and boundary conditions

Nodes: number of nodes

η: parameter

Type: choice between sinc or Lagrange interpolation function.

As an example, the Notebook illustrates the case of nonlinear boundary conditions and a biquadratic initial condition. The list of commands is given by

```
{Nodes,η}={21,0.1};
InitCond=4x^2(x-1)^2;
Bound0Cond=0;
Bound1Cond=0;
IBData={Bound0Cond,InitCond,Bound1Cond};
DiffNL[IBData, Nodes, η, Lagrange]
```

References

Arlotti L. and Bellomo N. (1995), On a new model of population dynamics with stochastic interaction, *Transport Theory Statistical Physics*, **24**, 431–443.

Arlotti L., Bellomo N. and Lachowicz M. (2000), Kinetic equations modelling population dynamics, *Transport Theory Statistical Physics*, **29**, 125–139.

Artioli E., Gould P. L. and Viola E. (2005), A differential quadrature method solution for shear-deformable shells of revolution, *Engineering Structures*, **27**, 1879–1892.

Artioli E. and Viola E. (2005), Static analysis of shear-deformable shells of revolution via G.D.Q. method, *Structural Engineering and Mechanics*, **19**, 459–475.

Back I. V., Blackwell B. and Saint Claire C. R. (1985), **Inverse Heat Conduction**, Wiley, New York.

Barenblatt G. I. (2003), **Scaling**, Cambridge University Press, Cambridge.

Bellman R. (1969), **Stability Theory in Differential Equations**, Dover Publications, Inc., New York.

Bellman R. and Casti J. (1971), Differential quadrature and long term integration, *J. Mathematical Analysis Applications*, **34**, 235–238.

Bellman R., Kashef B. and Casti J. (1972), Differential quadrature: Solution of nonlinear partial differential equations, *J. Computational Physics*, **10**, 40–52.

Bellomo N., De Angelis E., Graziano L. and Romano A. (2001), Solution of nonlinear problems in applied sciences by generalized collocation methods and Mathematica, *Computers Mathematics with Applications*, **41**, 1343–1363.

Bellomo N., Delitala M. and Coscia V. (2002), On the mathematical theory of vehicular traffic flow. I. Fluid dynamic and kinetic modelling. *Mathematical Models and Methods in Applied Sciences*, **12**, 1801–1843.

Bellomo N. and Flandoli F. (1988), Stochastic partial differential equations in continuum physics, *Mathematics Computers in Simulations*, **31**, 3–17.

Bellomo N. and Preziosi L. (1996), **Modelling, Mathematical Methods and Scientific Computation**, CRC Press, Boca Raton.

Bellomo N., Preziosi L. and Romano A. (2000), **Mechanics and Dynamical Systems with Mathematica**, Birkhäuser, Boston.

Bert C. and Malik M. (1996), Differential quadrature method in computational mechanics, *ASME Review*, **49**, 1–27.

Bert C. and Malik M. (1998), Three-dimensional elasticity solutions for free vibrations of rectangular plates by the differential quadrature method, *International Journal of Solids and Structures*, **35**, 299–318.

Bonzani I. (1997), Solution of nonlinear evolution problems by parallelized collocation-interpolation methods, *Computers Mathematics with Applications*, **34**, 71–79.

Bonzani I. and Mussone L. (2003), From experiments to hydrodynamic traffic flow models. I. Modelling and parameter identification, *Mathematical Computer Modelling*, **37**, 1435–1442.

Cattani C. and Rushchitsky J. (2007), **Wavelet and Wave Analysis as Applied to Material of Micro- and Nanostructure**, World Scientific, London, Singapore.

Chen C. N. (1999), Generalization of differential quadrature discretization, *Numerical Algorithms*, **22**, 167–182.

Chen C. N. (2000), Differential quadrature element analysis using extended differential quadrature, *Computers Mathematics with Applications*, **39**, 65–79.

COLTON D., EWING R. AND RUNDELL W., EDS. (1990), **Inverse Problems in Partial Differential Equations,** Proceedings of the conference held at Humboldt State University, Arcata, California, SIAM, Philadelphia.

CROSS M. C. AND HOHENBERG P. C. (1993), Pattern-formation outside of equilibrium, *Rev. Modern Phys.*, **65**, 851–1112.

DAUBECHIES I. (1992), **Ten Lectures on Wavelets**, SIAM, Philadelphia.

DAUTRAY R. AND LIONS J. L. (1990), **Mathematical Analysis and Numerical Methods for Science and Technology**, Vol. 1–9, Springer-Verlag, Berlin.

DODD R. K., EILBECK J. C., GIBBON J. D. AND MORRIS H. C. (1982), **Solitons and Nonlinear Wave Equations,** Academic Press, Inc., London-New York.

ELDER J. W. (1959), The dispersion of matter in turbulent flow through a pipe, *Proceedings Royal Society London A*, **219**, 186–203.

EVANS, L. C. (1998), **Partial Differential Equations,** American Mathematical Society, Providence.

FISCHER H. B., LIST E. J., KOH R. C. Y., IMBERGER J. AND BROOKS N. (1979), **Mixing in Inland and Coastal Water**, Academic Press, San Diego.

FOWLER A. C. (1997), **Mathematical Models in the Applied Sciences**, Cambridge Texts in Applied Mathematics, Cambridge University Press, Cambridge.

GOTTELIEB D. AND ORSZAG S. A. (1977), **Numerical Analysis of Spectral Methods: Theory and Applications**, SIAM, Philadelphia.

GRINDROD P. (1996), **The Theory and Applications of Reaction-Diffusion Equations. Patterns and Waves**, Oxford Applied Mathematics and Computing Science Series, Oxford University Press, New York.

GUPTA A. AND CVETKOVIC G. (2000), Temporal moment analysis of tracer discharge in streams: Combined effect of physiochemical mass transfer and morphology, *Water Resource Research*, **36**, 2985–2997.

HENDERSON F. M. (1966), **Open Channel Flow**, Prentice-Hall, Upper Saddle River.

HUNDSDORFER W. AND VERWER J. (2003), **Numerical Solution of Time-Dependent Advection-Diffusion-Reaction Equations**, Series in Computational Mathematics, **33**, Springer-Verlag, Berlin.

JAGER E. AND SEGEL L. A. (1992), On the distribution of dominance in populations of social organisms, *SIAM J. Applied Mathematics*, **52**, 1442–1468.

JOHNSON R. S. (1997), **A Modern Introduction to the Mathematical Theory of Water Waves**, Cambridge University Press, Cambridge.

KARAMI G. AND MALEKZADEH P. (2002), A new differential quadrature methodology for beam analysis and the associated differential quadrature element method, *Computational Methods Applied Mechanical Engineering*, **191**, 3509–3526.

KARAMI G. AND MALEKZADEH P. (2004), Vibration of non-uniform thick plates on elastic foundation by differential quadrature method, *Engineering Structures*, **26**, 1473–1482.

KAYA D. (2004), Exact and numerical soliton solutions of some nonlinear physical models, *Applied Mathematics Computation*, **152**, 551–560.

LAVRENT' EV M. M., REZNITSKAYA K. G. AND YAKHNO V. G. (1985), **One-Dimensional Inverse Problems of Mathematical Physics**, AMS Translations, Series 2, **130**, American Mathematical Society, Providence.

LEOPOLD L. B. AND MADDOCK T. JR. (1953), The hydraulic geometry of stream channels and some physiographic implications, *U.S. Geological Survey Professional Paper*, **252**.

LIN C. C. AND SEGEL L. A. (1988), **Mathematics Applied to Deterministic Problems in the Natural Sciences**, SIAM, Philadelphia.

LIONS J. L. AND MAGENES E. (1969), **Problèmes aux Limites Non Homogènes et Applications**, Vol. 1–3, Dunod, Paris.

LUND J. AND BOWERS K. (1992), **Sinc Methods**, SIAM, Philadelphia.

LYNCH S. (2007), **Dynamical Systems with Applications Using Mathematica**, Birkhäuser, Boston.

MARASCO A. AND ROMANO A. (2002), **Scientific Computing with Mathematica: Mathematical Problems for Ordinary Differential Equations**, Birkhäuser, Boston.

Meyer Y. and Ryan R. D. (1993), **Wavelets: Algorithms and Applications**, SIAM, Philadelphia.

Mickens R. E. (1996), **Oscillations in Planar Dynamical Systems**, World Scientific, London, Singapore.

Murray J. D. (1993), **Mathematical Biology**, Biomathematics, **19**, Springer-Verlag, Berlin.

Parkes E. J. and Duffy B. R. (1996), Travelling solitary wave solutions to a seventh-order generalized KdV equation, *Physics Letters A*, **214**, 271–272.

Revelli R. and Ridolfi L. (2005), Nonlinear convection dispersion models with a localized pollutant source. II. A class of inverse problems, *Mathematical Computer Modelling*, **42**, 601–612.

Romano A., Lancellotta R. and Marasco A. (2005), **Continuum Mechanics Using Mathematica**, Birkhäuser, Boston.

Satofuka A. (1983), A new explicit method for the solution of parabolic differential equations, in **Numerical Methodologies and Properties in Heat Transfer**, Hemisphere, New York, 97–108.

Schnakenberg J. (1979), Simple chemical reaction systems with limit cycle behaviour, *J. Theoretical Biology*, **81**, 389–400.

Schnoor J. L. (1996), **Environmental Modeling**, Wiley, New York.

Sewell G. (1988), **The Numerical Solution of Ordinary and Partial Differential Equations**, Academic Press, London-New York.

Shen S. (1993), **A Course on Nonlinear Waves**, Nonlinear Topics in the Mathematical Sciences, **3**, Kluwer Academic Publishers Group, Dordrecht.

Shu C. (1992), **Differential Quadrature and its Application in Engineering**, Springer-Verlag London, Ltd., London.

Shu C. and Richards B. E. (1992), Application of generalized differential quadrature to solve two-dimensional incompressible Navier-Stokes equations, *International Journal for Numerical Methods in Fluids*, **15**, 791–798.

Stenger F. (1983), Numerical methods based on Wittaker cardinal or Sinc functions, *SIAM Review*, **23**, 165–224.

Stenger F. (1993), **Numerical Methods Based on Sinc and Analytic Functions**, Springer, Berlin-New York.

Whitham G. (1974), **Linear and Nonlinear Waves**, Wiley, New York.

Subject Index

Ill-Posed Problems

Initial Value Problems

Mathematical Models

Mixed Problems

Neumann Problems

Ordinary Differential Equations